THE POCKET IDIOT'S GUIDE™ TO

The iPhone

by Damon Brown

ALPHA

A member of Penguin Group (USA) Inc.

ALPHA BOOKS

Published by the Penguin Group

Penguin Group (USA) Inc., 375 Hudson Street, New York, New York 10014, USA

Penguin Group (Canada), 90 Eglinton Avenue East, Suite 700, Toronto, Ontario M4P 2Y3, Canada (a division of Pearson Penguin Canada Inc.)

Penguin Books Ltd., 80 Strand, London WC2R 0RL, England

Penguin Ireland, 25 St. Stephen's Green, Dublin 2, Ireland (a division of Penguin Books Ltd.)

Penguin Group (Australia), 250 Camberwell Road, Camberwell, Victoria 3124, Australia (a division of Pearson Australia Group Pty. Ltd.)

Penguin Books India Pvt. Ltd., 11 Community Centre, Panchsheel Park, New Delhi—110 017, India

Penguin Group (NZ), 67 Apollo Drive, Rosedale, North Shore, Auckland 1311, New Zealand (a division of Pearson New Zealand Ltd.)

Penguin Books (South Africa) (Pty.) Ltd., 24 Sturdee Avenue, Rosebank, Johannesburg 2196, South Africa

Penguin Books Ltd., Registered Offices: 80 Strand, London WC2R 0RL, England

To Parul.

Contents

Introduction

There's usually a gap between when a new technology comes out and when it becomes a must-have. TiVo was out for months, if not years, before it really took off. The now-commonplace iPod didn't become a necessity until 2 years after its launch. Technology takes time. Well, it usually does.

The iPhone is the exception. Before it was even officially announced, rumors flew, speculating what features it would have, when Apple would officially announce it, what it would look like, etc. And as expected, it all hit fever pitch when Apple officially announced the iPhone in January 2007, making the phone the must-have device of the year—6 months before it was even available. Customers were undaunted by the high price tag, inevitable shortages, and predictable long lines. No wonder: the iPhone is an iPod, a portable Internet browser, a high-definition camera, a movie player, a worldwide phone, and more—all in one device.

Apple is known for making user-friendly devices, as your iPod-loving grandmother may attest, but the multifunctionality that makes the iPhone brilliant also makes it one complicated device. That's where this book comes in. In these pages, I help you navigate the coolest device of the new millennium. I show you …

- ◆ The quick and easy way to get started immediately.
- ◆ How to get music and movies into your phone.

◆ Tips for getting the best out of the camera,
Internet, music and movie players, and cell
phone features.

◆ And more!

The iPhone is a slick, powerful machine. And with
this book, deciding if it's worth the money won't be
a tough call.

Extras

Throughout the book, you'll see sidebars—handy,
bite-size pieces of information that offer definitions
of technical terms, warnings about possible challenges, and other tips and tricks to help you understand the iPhone. Here's what to look for:

> **iTerms** _____
>
> Learning all the iPhone lingo can be an
> iPain. iTerms sidebars clue you in on
> the latest techie jargon. Armed with this
> knowledge, you're bound to look cool in
> front of the geeks.

> **Incoming Call** _____
>
> Little tips and tricks will make your
> iPhone experience a lot smoother. Find
> them in these Incoming Call sidebars.

Crossed Signals

Crazy stuff is bound to happen anytime you're messing with a new device for the first time. Crossed Signals sidebars help you avoid potential hang-ups.

Music to Your Ears

Fun and interesting tidbits can give you insight into the little marvel that is the iPhone. That's what you'll find in Music to Your Ears sidebars.

Acknowledgments

To my support network, including my mother, Bernadette Johnson; my girlfriend, Dr. Parul Jashbhai Patel; and my mentor, Jane Briggs Bunting—it couldn't have happened without you.

This book was a group effort, not a solo one, and I need to thank my agent, Marilyn Allen, and my acquisitions editor, Michele Wells, for their unwavering faith. Third time's the charm!

Finally, thank you Apple Corporation, for the product.

Trademarks

All terms mentioned in this book that are known to be or are suspected of being trademarks or service marks have been appropriately capitalized. Alpha Books and Penguin Group (USA) Inc. cannot attest to the accuracy of this information. Use of a term in this book should not be regarded as affecting the validity of any trademark or service mark.

The iPhone: Good Call!

In This Chapter

- Why the iPhone is unique
- What model iPhone is for you?
- The costs and considerations of upgrading to an iPhone

Today, several cell phone models can carry and play music, but the iPhone is the only one that's compatible with Apple's iTunes—plus, it's so much cooler than other phones! In this chapter, you get the ins and outs of the iPhone, a comparison of the models available, and an idea of how much your leap into the next generation of phone technology is going to cost you.

What Is the iPhone?

The iPhone is Apple's answer to the *smartphone*, a multimedia cell phone. Unlike most cell phones on the market today, the iPhone can play your favorite music and videos as well as provide full Internet and e-mail access.

iTerms

A **smartphone** is any phone with functions similar to a personal computer.

Perhaps the coolest feature of the iPhone is that there are no keys or buttons to push. To operate and navigate it, you use your finger to "press" the different icons on the iPhone's touch screen, similar to PDAs or some of the high-tech computers you may have seen at office buildings, ATMs, or the supermarket.

Why Is the iPhone Better Than Other Cell Phones?

Why *isn't* it? Until recently, most cell phones had one primary purpose: making and receiving calls. Every other function was something of an afterthought. The iPhone is one of the first phones to integrate a music player, a movie player, a laptop, and even a digital camera into one device.

Check out the following table, which shows some more of the iPhone's features compared with traditional cell phones.

Traditional Cell Phones vs. the iPhone

	Traditional Cell	iPhone
Buttons	keys	touch screen
Screen size	about 1.5 inches	3.5-inch widescreen
Camera	varies	2.0 megapixels (MP)
Internet	limited, if any	built-in Wi-Fi
E-mail	limited, if any	all major carriers
Worldwide operation	not likely	in most countries
Battery life	about 1 day of use	about 5 hours of use
Cost	less than $100	starts at $499

Music to Your Ears

Apple expects to sell at least 10 million iPhones within a year of its June 2007 release. And that's *not* a far-fetched estimate, if iPod sales are any indication: Apple sold more than 21 million iPods the preceding holiday season.

The iPhone also has several functions most cell phones don't have. For instance, Safari is a fully functional web browser that allows you to surf the web from anywhere (anywhere you have *Wi-Fi* access, that is). The included Bluetooth capabilities also allow you to transfer files wirelessly from other Bluetooth-enabled devices.

iTerms

Wi-Fi is any Internet connection that allows users to connect wirelessly.

iPhone Models

The iPhone comes in two models, differentiated only by memory size: 4 gigabytes (GB) or 8 GB. The more memory you pony up for, the more music and videos you can store on the device. Aside from price and memory, both models are alike. The following table lists some quick specs of the two models.

iPhone Models

	4 GB Model	8 GB Model
List price	$499.00	$599.00
Songs (approximately)	1,000	2,000
Camera	2 MP	2 MP
Weight	less than 5 oz.	less than 5 oz.
Dimensions	4.5×2.5×.5 inches	4.5×2.5×.5 inches

iPhone Considerations

On January 9, 2007, the day of the official iPhone announcement, millions of techies (and plenty of nontechies) were chomping at the bit to get the hottest new toy from Apple—even with the $500 or $600 price tag. If you've got your heart set on getting an iPhone, or if you have one on order, obviously I'm not going to talk you out of getting one. (You wouldn't need this book then!) But I do have to bring up some commonsense iPhone-related considerations you should think about.

New Toys Don't Come Cheap

As I've mentioned, the iPhone does have a hefty price tag. Even the cheapest iPhone, with tax, costs you more than $500. If you said to yourself, *Yeah, so what? I'll spend that*, keep reading.

What cell phone carrier do you have? As of summer 2007, only Cingular customers are being offered the iPhone. This means that, yes, you can go to your local Apple store and buy an iPhone or purchase one online at www.apple.com, but technically, the only way you can get service is if you are or become a Cingular customer.

This means that if you're a customer with Sprint, T-Mobile, or any other wireless carrier, you'll have to break your current contract and sign up with Cingular to use the iPhone. Of course, those with the willpower to set aside this beautiful, shiny phone can patiently wait for their current contract to expire. But most people won't.

> **Incoming Call**
>
> As of summer 2007, only Cingular customers can use the iPhone, although Apple is expected to include other major carriers in the future. So don't cancel that non-Cingular contract just yet—things might soon change in your favor! Visit www. apple.com to get the latest updates.

Apple and Cingular currently require a 2-year contract—double the standard 1-year contract—for iPhone customers. As of this writing, the actual monthly fee for the iPhone is yet to be determined. However, with the iPhone's BlackBerry-like functionality, it's safe to assume the monthly charge will be in the $70 or $80 range.

I'm not trying to talk you out of buying an iPhone, but as you can imagine, the costs can add up:

- ◆ The cheapest iPhone retails for $499.
- ◆ Breaking a current cell phone contract can cost up to $200 in penalty fees.
- ◆ Possibly doubling your cell phone bill adds up to hundreds of extra dollars a year.

But do keep in mind that the prices of tech products usually drop after the first year. So if you can hold out, you might see the benefits in your wallet.

Crossed Signals

If you're fortunate enough to buy an iPhone early, you're getting what's called, in tech talk, the "first-generation model," which comes standard with a higher likelihood of kinks, challenges, and crashes than any later versions. (It sometimes takes a version or two for the manufacturer to work out the bugs.) It's wonderful to be on the cutting edge, but understand the risks you're taking.

iPhone + Computer = Music to Your Ears

To take full advantage of all the iPhone has to offer, you need to have a computer. Here are the requirements for Mac users:

- ◆ USB 2.0 port
- ◆ OS X operating system, v10.4.8 or higher
- ◆ Internet access
- ◆ iTunes 7 or higher

And for PC users:

- ◆ USB 2.0 port
- ◆ Windows 2000 (Service Pack 4), or Windows XP Home or Professional (Service Pack 2), or higher
- ◆ Internet access
- ◆ iTunes 7 or higher

Even if you don't have an iPod and don't plan on getting one, still get iTunes. You can download the latest version for free at www.apple.com/itunes/download.

I discuss more of what you need your computer and iTunes for in later chapters, so stick with me here.

iPhone vs. Other Electronic Gadgets

Okay, you might be thinking, *the iPhone is perfect! It's got my camera for when I travel, the Internet for browsing, e-mail for communicating, movies for long trips, and music for jamming. And I can also make phone calls!*

The iPhone's camera is great when you have an unexpected photo op, but don't expect to take high-quality pictures like you would with your digital camera. At 2 MP, the visual quality of iPhone pictures won't match your digital camera, which probably clocks in at around 3 MP.

Will you use your iPod as much after you get your iPhone? It depends. As I noted earlier, the iPhone can hold 1,000 (4 GB model) to 2,000 (8 GB model) songs. If that's enough for you, great! If not, don't sell your iPod on eBay just yet.

Although the iPhone can't beat traditional digital cameras, laptops, or iPods, it is much better than the average multifunctional cell phone. But did I even have to tell you that?

Music to Your Ears

They say imitation is the highest form of flattery, and some of Apple's competitors are proving that by releasing iPhone-like devices of their own. Two such iClones are the LG Prada (www.lgpradaphone.com), the LG phone officially sponsored by the high-end couture label Prada, and Motorola's MOTOMING (direct.motorola.com/hellomoto/motomingedge), which comes with a bonus FM radio.

If you're convinced an iPhone is for you, or if you've already taken the plunge and have your iPhone, turn to Chapter 2 to learn how to get started using it.

The Least You Need to Know

◆ With the iPhone, you can make calls, take pictures, watch movies, play music, browse the web, and send e-mails—all with one cool device.

◆ The 4 GB iPhone costs $499 and holds 1,000 songs, while the 8 GB iPhone costs $599 and holds 2,000 songs.

◆ The iPhone sports a slick, touch screen instead of keys or buttons.

◆ Does your cell phone carrier offer the iPhone? If it doesn't right now, you can either cancel your contract and switch to a carrier that does, or wait until Apple signs up more carriers.

Getting Started

In This Chapter

- ◆ A tour of your new iPhone
- ◆ A look at what's on the iPhone menus
- ◆ Getting touchy with the touch screen
- ◆ Making that first call

For all its different functions, the iPhone definitely packs a lot of punch in such a small device. And at first glance, you'll see it has relatively few actual buttons on the surface. So how do you turn it on? How do you navigate it? How do you make a simple phone call or retrieve your voicemail?

That's what this chapter is all about. In the following pages, I give you the ins and outs of getting started with the iPhone. As with most Apple products, it's very user friendly, but having previous experience with the iPod helps.

Examining Your iPhone

I'm sure you're probably chomping at the bit to get into your new iPhone, make a call, play some music, or send an e-mail. But not so fast. Let's take a minute to explore the outside of the iPhone. If you have your iPhone handy, pick it up and let's go on a quick tour. Let's start with the back of the iPhone and save the all-important front for last.

The Back

The back of your iPhone has only one item: the camera. With the iPhone's 2 megapixel (MP) camera, you can shoot photos, store them in your iPhone, or send them to your friends. As I noted in Chapter 1, the 2 MP camera is fine as far as cell phone cameras go, but don't expect to get professional-quality photos like you'd get on your full-fledged, 3 MP digital camera. Still, the camera is a cool feature on the phone.

 Incoming Call _____

The iPhone makes an excellent back-up camera for when you have an unexpected photo op ... but your digital camera is sitting at home.

The Side

Rotate around to the left side of your iPhone. Here you'll find the ringer button and the volume control.

Like other cell phones, you simply press the ringer button to turn on or silence the phone ring. This quick button comes in handy when you're in a meeting, a concert, or a movie theater.

Right below the ringer button is the volume control. Push this button to increase the sound coming out of the iPhone, both the speaker/headset volume and the volume of your friend's voice when you make a phone call.

The Top

Turn your attention to the top of your iPhone. Here you find three items:

- 3 mm headset jack
- SIM card slot
- Sleep/Wake button

The headset jack is pretty self-explanatory. You plug in your headphones here to listen to your iPhone jams privately.

Music to Your Ears

The iPhone comes with the standard white iPod "ear buds," but you can use your own headphones, too. Any headphones that would plug into your standard iPod, CD player, or tape player work with your iPhone.

The SIM card, nestled at the top of your iPhone, holds all the phone numbers, contact information, and other preferences for your device. All cell phones use SIM cards. Like a portable disc or hard drive, SIM cards make it easy to retain your information if your cell phone dies. As long as the card is intact, you can simply take out the SIM card and put it into another phone of the same model.

The Sleep/Wake button temporarily powers down your iPhone without turning it off, so if you have to stop in the middle of something, you won't lose what you were working on. Say you're writing an e-mail on the iPhone, but you have to stop to head to a meeting. You can hit the Sleep/Wake button to make your iPhone "sleep," attend your other business, hit the button again to "awaken" the phone, and resume your e-mail right where you left off. This is similar to closing the lid of your laptop. All your information is still there; the machine just goes on hold for a bit.

Crossed Signals

The iPhone uses much less energy while in sleep mode, but it still uses battery power. Turn your phone completely off if you need to conserve every ounce of power.

The Bottom

Now flip the iPhone and look at the bottom. You'll see these three items there:

- ◆ Speaker
- ◆ iPod connector
- ◆ Mic

The speaker plays all the sounds and music coming from your iPhone. Plugging in some headphones redirects the sound and keeps your music private.

The iPod connector is the same hole you'll find at the bottom of your iPod. In fact, the iPhone uses the same cord the iPod uses to connect to a computer. The USB connector on the other end of the cord plugs into your computer and allows your iPhone and computer to "talk" and sync up.

The mic(rophone) is, you guessed it, where you talk into the device. No, you don't have to hold the phone upside down to talk into it for a phone conversation; the mic is powerful enough to pick up your voice as you hold it normally.

The Front

Now, finally, for the front of your iPhone, where the magic happens. There are three items on the front of your iPhone:

- ◆ Ear hole
- ◆ Touch screen
- ◆ Home button

The ear hole is a little slit that allows you to hear conversations—just like the earpiece on any other phone. Nothing fancy there.

The touch screen—talk about fancy—is what you use to interact with your iPhone. It is 3.5 inches (diagonally) and 160 *ppi* (*pixels per inch*), meaning it has one of the largest, sharpest resolutions available on a portable device of its size.

> **iTerms** _____
>
> **PPI** stands for **pixels per inch,** the measure of how detailed a particular camera's photos are. The higher the pixels per inch, the more memory the picture requires.

Why is it a touch screen instead of a regular cell phone screen? Because this monitor serves as both your screen and your keyboard. You use your index finger (or thumb, if you like) to "press" the highlighted buttons onscreen. With the iPhone, your fingers essentially replace the mouse and keyboard. (More on the touch screen coming up.)

Music to Your Ears _____

Touch screen technology has been around for years, used in PDAs (personal data assistants), computer screens, and other applications. Cell phone manufacturers didn't start implementing it until late 2006.

Finally, the appropriately named Home button takes you back to the default iPhone home menu screen, the screen that pops up when you turn on the phone. This comes in handy if you get lost in the menus.

Starting Your iPhone

Some of you probably figured out how to turn on the phone almost before you had it out of the box, but here's how to do it for the rest of you: when you first turn on the iPhone, you're greeted with a hello message and a basic screen with the time and date. At the bottom you'll see a button and the message "slide to unlock." Put your finger on the button, and slide it along the slot to the right. Your iPhone is now unlocked and ready for you to use it. Now for the cool stuff.

reception

time and date

battery life

unlock button

To unlock and start your iPhone, you simply slide the arrow. Like many other Apple products, it's very user friendly.

Using the Touch Screen

You've probably seen someone on the train or in a meeting, "writing" on their PDA or smartphone touch screens with a stylus (a little inkless "pen"). But those styluses are easy to lose, and you have an extra step of getting the thing out anyway.

Not so with the iPhone. You only need a few tools to use the iPhone's *touch screen*—and they're attached to the end of your hand so you'll never lose them. These tools? Your index finger and your thumb. You truly let your fingers do the walking with the iPhone, with its "multi-touch display"—Apple's name for the iPhone's touch screen. That should be easy to use, right? Well, it is, but you do have to do things in a way it will understand. Here's how:

Touch Screen Tips	
Scroll up	Lightly put your finger on the screen and push toward the bottom.
Scroll down	Lightly put your finger on the screen and push toward the top.
Select	Push the button or desired item.
Expand	Place your thumb and index finger on the border of the item and pull your fingers apart.
Shrink	Place your thumb and index finger on the border of the item and pull your fingers together.

iTerms

A **touch screen** is a computer monitor that reacts based on when and where you touch it.

The iPhone Menus

Now that you know how to navigate your iPhone's touch screen, let's look at the menus you have to work with. The iPhone offers several options from its default home menu screen.

iPhone Home Menu (Top)

Reception	The clarity of your calls, which can vary based on your location
Text	Send a quick text message to another phone
Calendar	Check the date
Photos	Shuffle through your photo collection
Camera	Take a picture
Calculator	Do some math
Stocks	Check your stock portfolio
Maps	Chart a destination
Weather	Look at the weather
Notes	Write quick notes to yourself
Clock	Check the time
Settings	Change your iPhone setup

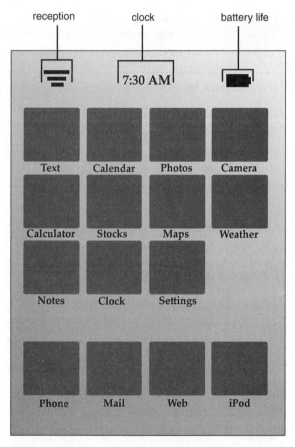

Use the iPhone's home menu to navigate quickly and easily.

At the bottom of the screen you'll see some more buttons. These take you to the various parts of your iPhone:

iPhone Home Menu (Bottom)	
Phone	Make a call, look up contacts, etc.
Mail	Send and receive e-mails
Web	Surf the web
iPod	Listen to and/or organize your music

(I give more details on the different menu functions in later chapters.)

iPhone *Phone* Basics

At the end of the day, the iPhone, despite all its high-tech gadgetry, is still a phone, and making that first, "Hey, I'm calling on my new iPhone!" call is a special occasion. The following sections show you how.

Turning On Service

First, you'll need cellular service. If you got your iPhone at a Cingular store or other authorized cell phone carrier, you might already have service set up. But if you got your phone at an Apple store or at www.apple.com, you might need to arrange to get your cellular service set up and turned on yourself. Like most other cell phones, there are four steps to activating your iPhone:

1. Call Cingular or another iPhone-supporting carrier.

2. Choose the best iPhone package for your needs.

3. Listen to and agree with the terms.

4. Make the purchase.

Crossed Signals

As of summer 2007, Cingular was the only carrier to support the iPhone. Others will probably support it later, but don't buy the phone now and assume your carrier will eventually be added to Apple's carrier list. You might have to break your contract if it comes to that.

Making a Call

Now for that all-important first call. From the bottom of the home menu, press the phone icon on the bottom left. On the next screen you'll find five menu options.

---

iPhone Phone Menu

Favorites	A list of the calls you make most often
Recents	Your most recent calls made and received
Contacts	A list of your phone contacts
Keypad	Pulls up a traditional number pad to dial
Voicemail	Your messages

Press the Keypad icon, and a traditional phone number dial pops up. Type in the number you want to dial and press the green "Call" button.

After you have your contacts set up, you can press the Contacts icon to shuffle through your Contacts list. Simply click on a person's name, select the appropriate number (home, work, or mobile), and ring them up. No more punching in numbers on a keypad.

Music to Your Ears _____

In addition to phone numbers, you can store e-mail addresses, snail mail addresses, photos, and other information for your contacts. If you have a Mac, all this info syncs up from Address Book.

Receiving a Call

When news gets out you got a new iPhone, the calls will start coming in. Here's how to answer them.

When you receive an incoming call, you'll see the person's name at the top of the screen (if they're already stored in your Contacts list; otherwise, you might just see the phone number). At the bottom is a red button labeled "Don't Accept Call" and a green button labeled "Accept Call." Press the red button to send the call directly to voicemail. Press the green one to start the call.

Checking Your Voicemail

When you pressed the Phone icon, you saw five different options on the screen. The last option on the right is Voicemail. (If necessary, hit the Home button to get back to the top menu, press the Phone icon, and press the Voicemail icon.)

One of iPhone's many unique features is how it treats voicemail like e-mail—Apple calls it "visual voicemail." When you press the Voicemail button, you'll see a list of your recently received as well as your saved voicemails. Use your finger to scan through the list and then press the voicemail you want to listen to. This is a feature new to the iPhone; with other phones, you have to listen to the voicemails in order and you often can't see who the voicemails are from.

Crossed Signals

You're not going to have any voice-mails if you just bought the phone, so don't worry if it's initially blank.

Managing Your Contacts

The Contacts list is the first page you see in the phone setup. (You can get back here by pressing the Contacts button at the bottom of the phone menu.) Like other cell phones, managing your contacts makes staying in touch with your friends, family, and co-workers much easier.

Your contacts are listed alphabetically from top to bottom. Use your index finger to scroll along the listing until you see the person you want to call and then simply click on his or her name.

Notice the small plus button in the upper-right corner? Press that button to add a new contact. You can then type in their name and phone number(s), as well as e-mail address, home address, and other information.

You can also transfer contacts from your Mac or PC, but the method depends on your system and setup. Visit www.apple.com for more details on transferring contacts.

The Least You Need to Know

- ◆ The aptly named Home button (on the front of your iPhone, just below the touch screen) returns you to the main, or home, menu.

- ◆ Apple's multi-touch display enables you to navigate by gently pressing the screen and pushing your finger up or down to move around the menus.

- ◆ Before you can use your new iPhone, you need to have cellular service. As of summer 2007, Cingular was the only iPhone-supporting carrier.

- ◆ Making and receiving calls is easy with the iPhone's user-friendly interface and Contacts list.

iTunes

In This Chapter

- ◆ Locating and playing your music
- ◆ Creating your own playlists
- ◆ Sharing your playlists
- ◆ Tuning in to iTunes Radio

iTunes is the virtual jukebox that plays all your music (and videos—but that's another chapter). You can also use it to listen to podcasts and audiobooks. Basically, everything you want to do with your iPhone or iPod, you do through this handy interface. iTunes also enables you to pick up radio stations, making the program as sophisticated as the stereo sitting on your entertainment center shelf.

Play Me

Let's take a look at the basic iTunes screen:

play and volume options

song info

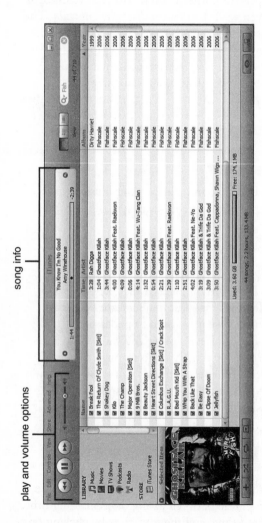

The iTunes interface.

The upper-left corner is similar to your traditional CD player, with the play and volume controls. The large forward arrow plays your current selection. (You can also double-click on a song to play it. You can still use the play icons to control the music.) When the music plays, the arrow turns into a double vertical bar, which you can use to pause the selection playing. The double forward arrow skips to the next selection, and the double back arrow restarts or, with a double-click, goes to the previous selection.

The long bar to the right of these icons controls the volume. Drag the circle along the line to adjust it.

Incoming Call

Later editions of iTunes play music videos and movies, too. You control videos the same way as the music. See Chapter 6 for more.

What Is This Song?

The panel to the right of the play icons, in the center of the screen, gives you information about the song currently playing. The top lines cycle between showing the song title, the album name, and the artist name.

Below the song info, iTunes shows the song time. You can shuffle between listing the remaining time, the elapsed time, and the total time by clicking on the information. For a quick glance, the bar shows visually how much is left to play. Like with the volume control, you can slide the diamond along the bar to move to different parts of the song. The little triangle to the left of the song information switches the face plate between song info and a cool visual equalizer.

At a Glance: Icons

Along the bottom of your iTunes screen you'll see several icons. From creating playlists to shuffling songs, these are quick ways to tailor your musical experience.

Here's what the icons do, from left to right:

iTunes Icons	
Plus/wheel	Create a playlist
Twisting arrows	Switches shuffle on and off
Circling arrows	Repeat song list, repeat song, or no repeat
Downward arrow	Show or hide album art
Box with arrow	Show video full screen (if applicable)
Browse	Browse and search through your music collection
Eject	Eject the CD, iPhone, or iPod

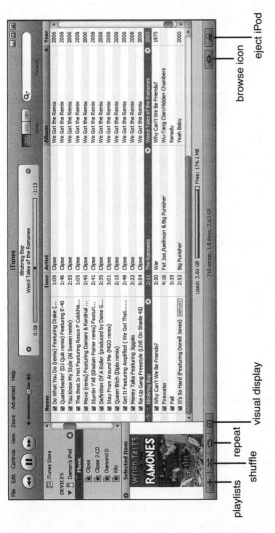

iTunes has several icons for quick navigation.

Incoming Call _____

You may see a "Turn on MiniStore" icon at the bottom of your iTunes player. Check this out to see what music Apple recommends based on your past purchases. Read more about it in Chapter 5.

Song 411

If it's in your library, you probably know something about the song already, but to check out detailed information, or to add some yourself, select a particular song by clicking on it. Moving right from the song title (on the left), you'll see all the pertinent song information:

iTunes Song Information	
Name	Title of song
Time	Length of song
Artist	Performer
Album	Title of the album/CD
Genre	Kind of music
My Rating	Your song rating (if any)
Play Count	How many times the song has been played
Last Played	Date and time of the last play

You can right-click on the column headings to pull up more song information options such as Year Released and Date Added To Catalog. You can also remove columns as you see fit.

For more detailed song information, you can use the Get Info option. Highlight the song title, right-click, and select the Get Info option. Alternatively, you can click on the File menu at the very top and select Get Info.

Along with the basic information, Get Info shows you the location of the song file on your computer's hard drive, composer data, and other information. You can also increase or decrease the volume of that particular song if the CD burned was too loud or soft. In later versions of iTunes, you can even add the lyrics for the song so you can sing right along as it plays.

Music to Your Ears

The program that looks up song information when you insert a CD, Gracenote, alphabetizes your music by the artist's first name. To organize by last name, put in the CD, highlight all the songs, right-click for Get Info, and change the artist name to last name, first name.

Rating Your Music

It's your favorite song. Every time you hear it, you're transported back in time to when you first heard it, danced to it, or that great day you drove around with the windows down while jamming to it. It gets five out of five stars in your book—and now in your iTunes, too, thanks to iTunes's rating system.

Here's how you rate a song:

1. Find the song in your library.
2. Move your cursor right along the song information row until you reach the My Rating column (between Genre and Play Count).
3. Click on the My Rating column, and you'll see five dots. Click the dot that corresponds to the rating you want to give—five dots/stars is the highest rating.

Rating music allows you to organize tracks better. Click on the Ratings heading at the top of your song columns in iTunes, and it lists your music from the highest rated to the lowest rated. Click it again to reverse the order. Nonrated songs are always ranked below the lowest rated.

Incoming Call _____

You can also assign a rating in Get Info. See the "Song 411" section for details.

Get the Party (Shuffle) Started

With Party Shuffle, you can set iTunes to play a random list of songs from your library or from specific playlists. Party Shuffle sets iTunes to replay so you never have to restart the music, leaving you free to go enjoy the party … or go clean the house (my editor's favorite use for Party Shuffle).

Start by clicking on the Party Shuffle icon in the Playlists section of the left column. A list of songs will pop up. At the bottom of the list of songs are different criteria you can use to modify your list. Pull down the Source menu to make a random list from songs other than your library. For instance, if you've made a playlist with all your favorite slow jazz songs called "Slow Jazz 1," you can scroll down the Source menu and tell iTunes to play random songs from that list.

Adjust the Display options to show the most recently played songs and the upcoming songs. You can check the box below the Source list to make iTunes play your higher-rated songs more often.

Finding Music

You've had a song stuck in your head for days now, and you can think of the title, but you're not sure of the artist. But you need to listen to the darn thing to get it out of your head. How do you find it in your iTunes library? You have a few options:

◆ Music search
◆ Browse
◆ Manual search

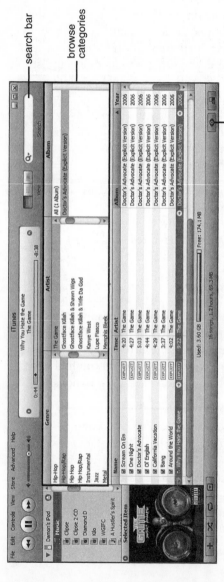

When you need to find music in your iTunes library, you have a few options.

ISO Music

The Search bar is located in the upper-right corner of your iTunes screen. Type in certain words (song title, author, album name), and iTunes lists the songs that fit that criteria. For instance, if you only want Billy Joel songs, click on the Search bar, type in "Billy Joel," and presto, you'll get a list of all Billy Joel–related songs. Press the small X on the right of the Search bar to end the search.

The Search bar looks for words through all the song information, meaning it not only lists songs for which Billy Joel is artist, but also any songs that list him as composer, writer, or in any other capacity. It also means that if some random group had a song title that contained the words *Billy Joel*, it would pop up on the list, too. For more control, click on the magnifying glass symbol and specify if you'd like to use the criteria to search for artists, albums, composers, songs, or all variables. Any more detail requires using the Browse function, coming up next.

Thanks, I'm Just Browsing

In the lower-right corner of your iTunes screen, look for the little eye icon. This is the Browse icon. Click on it, and iTunes opens up a new panel underneath the song information. Here, all your music is organized by Genre, Artist, and Album. Scroll through the various lists, and click on specific criteria to find what you want.

For instance, if you're in the mood for Alternative music, click on Alternative in the Genre box. iTunes lists all the music labeled as Alternative, and you can scroll through the other columns to find a specific artist or album within that genre. Clicking on the All category at the top of each list removes the selected criteria and expands your search. You can click the Browse eye again to close the browsing function.

There are three ways to view your music:

◆ By list

◆ Grouped by artist

◆ Cover Flow

List, the default setting, shows all your music in a traditional, text-only list.

Grouped by artist gives you another text listing but organizes everything by group or performer.

The very cool Cover Flow option gives you a jukeboxlike setting. All the album covers appear side by side. Flip through the music by dragging the sliding button at the bottom. Although visually different, everything functions the same in Cover Flow mode.

Incoming Call

Cover Flow is a really cool option Apple recently introduced. See Chapter 4 for more on Cover Flow.

You can switch your view by clicking on the view icons located just right of the music information at the top of iTunes.

Manual Search

iTunes lists all your music by song name, time, artist name, and album name. Alphabetically by title is the default library setup, but you can click on the column labels to organize your music the way you want. Then you can scroll up or down and find your songs.

> **Music to Your Ears**
>
> Clicking on an already highlighted column label puts the music in the reverse order—Z to A in the Name, Artist, or Album column and from longest to shortest in the Time column. Click it again to return the list to descending order.

Playing with Playlists

For those of you old enough to remember making mixed tapes (yes, tapes—they were around before CDs), welcome to the modern version of mixed tapes: playlists. With playlists, you can group your favorite songs under one heading. For instance, a playlist you name "Exercise Mix 1" may have all your favorite jogging songs. Drop that in your iPhone or iPod, plug in your headphones, and hit the road.

Standard Playlists

Some playlists come standard with iTunes:

- ◆ Recently Played
- ◆ Purchased
- ◆ Recently Added

Recently Played shows a list of the music you've just played. This is helpful if you have a large music collection and hear a song but don't catch the name of it.

Purchased gives you all the music you've bought through the iTunes store since starting your account.

Crossed Signals

The Purchased playlist resets if you have to reinstall iTunes or change your library to another computer.

Recently Added has the songs just added to your collection. These can be iTunes Store purchases or from CDs you imported into iTunes.

The Basic Playlist

It's easy to create a new playlist. Under the File menu, click New Playlist. A new playlist called "untitled playlist" appears in the left column, below Library, Party Shuffle, and other iTunes items, ready for you to type in a name. Don't worry—if you decide to change the playlist title later, you can double-click on the name and change it.

Use playlists to make your own music mixes.

Now click on Library (or another playlist, if that's where you plan on getting the music from) and find the first song you'd like to add to your new playlist. Put your cursor over the song, hold down the mouse button, and manually drag the song to where your new playlist is listed in the left column. Then just repeat the process until your playlist is complete. Remember that you're not dealing with CDs or another very limited medium, so your playlist can hold as many songs as your hard drive can.

> **Incoming Call**
>
> Your playlist song order isn't set in stone. Just pick a song you want to move, drag it up or down your list of playlist songs, and drop it where you like.

That's Smart!

Manually creating basic playlists can be a lot of work. Luckily, Apple thought of that and created Smart Playlists, which essentially do the work for you.

Under the File menu, select New Smart Playlist .... In the info box that appears, you can tell iTunes what criteria you'd like it to use create your new playlist based on item, relationship, and description. For instance, you can do Artist (item) contains (relationship) "Billy Joel" (description). The smart playlist then creates a playlist featuring all songs listing Billy Joel as the artist. Unlike the Search function, the smart playlist won't include songs for which Billy Joel is listed as composer or any other position other than artist.

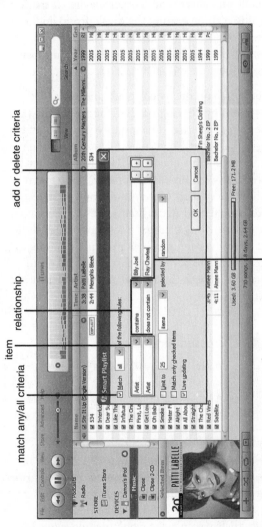

match any/all criteria

item

relationship

add or delete criteria

description

Let Smart Playlists create mixes for you.

Alternatively, if you want everything *but* songs that feature Billy Joel, you can do Artist (item) does not contain (relationship) "Billy Joel" (description). The result will be a playlist with all the songs not featuring Billy Joel.

To get more specific, you can add more playlist descriptions by clicking on the + button to the right of the initial criterion. For instance, if you want a playlist of all your Billy Joel songs, but you don't want to include his duet with Ray Charles, your first request would read Artist (item) contains (relationship) "Billy Joel" (description), and your second request would read Artist (item) does not contain (relationship) "Ray Charles" (description), as shown in the preceding figure. With multiple requests, you can tell iTunes to add songs that match all or any of the criteria; choose this option at the top left of the Smart Playlist box.

Crossed Signals

> If nothing's showing up in your Smart Playlist, double-check your All box—you could have two contradictory sets of criteria that no song matches!

Like the basic playlist, your Smart Playlist can be as long as you like. However, you can't control the order of the songs listed in it, only the criteria of the songs chosen.

Burning Your Playlists to CD

You put together a killer playlist, and you can't
wait to share it with all your friends. That's easy to
do with iTunes by burning your playlists to CD.
With a blank CD in your computer, highlight your
playlist and click the Burn Disc button in the lower-
right corner (next to the Browse icon). After a few
moments, during which you can chart the burning
process at the top center of the screen, iTunes will
have a copy of your playlist on CD, ready for you
to share.

Burning Tips

CD-R or *CD-RW* CDs work best for burning playlists, and packs with multiple CDs are now widely
available. The average CD holds 78 minutes of
music; if you have more than that in your playlist,
you'll have to create a shorter list. The total playlist time is shown at the bottom of iTunes when the
playlist is highlighted.

iTerms

CD-R (CD-recordable) and **CD-RW**
(CD-rewritable) are two formats of
recordable compact discs. CD-Rs can only
be recorded on once, while CD-RWs can
be erased and rewritten over again and
again.

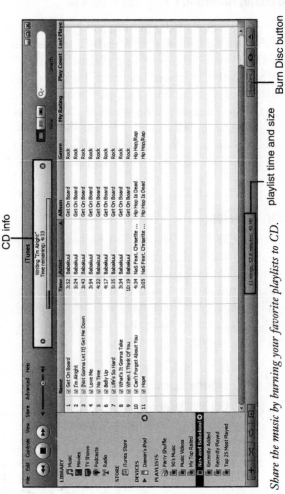

Share the music by burning your favorite playlists to CD.

Getting Album Art

The newest versions of iTunes grab artwork for you. Go to the Advanced menu and click on Get Album Artwork. iTunes then searches for the album art. You need an active Internet connection to get it.

When iTunes has your album art stored, it appears in the lower-left corner whenever you select a song from the CD.

> **Music to Your Ears**
>
> iTunes automatically gives you album art when you purchase a song or album from the iTunes Store.

To get album art ...

1. Get online, and using a search engine such as Yahoo! or Google, type in the album name.
2. Find a website with the album art, right-click on the picture, and choose "Save Picture As."
3. Save the picture on your computer, in the same area as your music files if possible.
4. Open iTunes and *downsize* the iTunes window so it only takes up part of the screen.
5. Find the album art on your computer, and click and drag it into the lower-left corner of iTunes where it says "Drag Album Artwork Here."

Album View highlights the album covers.

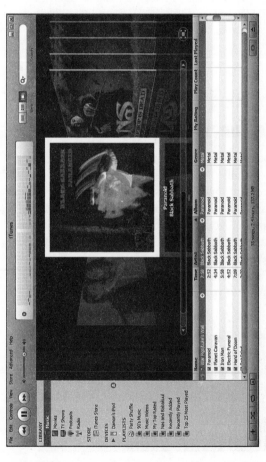

Flip through the album covers in Cover Flow View.

iTerms _____

Downsize means to make a particular window smaller. Aim for the lower-right corner, hold down the mouse, and adjust the window to the size you like.

Importing Previously Saved Music into iTunes

If you already have music on your computer, but it's not officially part of your iTunes library, get it in there! First, open iTunes and downsize it so you can still see part of your computer desktop. Next, find the area on your computer where your music files are stored. Highlight the music you want to transfer and click and drag the files into iTunes. iTunes will take a second or two to read each music file before adding it to your library.

Crossed Signals _____

Don't delete the original music file after adding it to iTunes! iTunes still needs the file to play the song.

Tune In: iTunes Radio

iTunes Radio connects you to online music stations around the world. Imagine picking up frequencies from Japan, Brazil, and the UK! It's also a great alternative way of getting hard-to-reach, streaming stations when you don't have direct web browser access.

To use iTunes Radio, you must first be connected to the Internet. Click on the Radio icon in the left column, and a list of available genres will appear, such as Blues, Classical, and Hard Rock.

Clicking on a particular genre unfolds the list of radio stations available. Each station lists its program title, sound quality, and program description. Double-click the title or highlight the program and press the Play button to start the music. It usually takes a second or two to start up because the music must be downloaded. When it begins, the current song information is listed in the face panel along with the radio station's web address.

Crossed Signals

Programs sometimes are temporarily offline, so move on to another program if the music hasn't started within a minute or so.

Regularly press the Refresh icon, located in the upper-right corner, to see if any new stations have become available since you started your iTunes session.

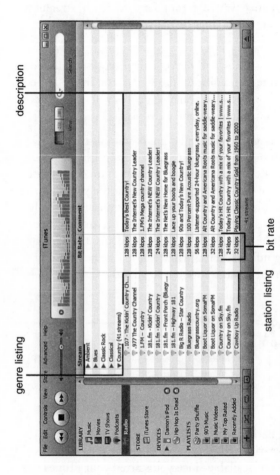

iTunes Radio picks up stations from around the globe.

The Least You Need to Know

◆ The user-friendly iTunes music controls are similar to your stereo or CD player.

◆ iTunes helps you stay organized and find songs and albums quickly with just a few mouse clicks.

◆ Playlists are homemade mixes you create. Think modern mixed tapes.

◆ Burn your playlists to CD with iTunes. Share with your friends. Enjoy.

Chapter 4

"Touch Your Music"

In This Chapter

- ◆ Getting music on your computer
- ◆ Filling your iPhone with music
- ◆ Browsing your music on your iPhone
- ◆ Organizing your digital music collection

"Touch your music."

Steve Jobs, Apple CEO, hit the nail on the head with this quote made during his January 9, 2007, keynote address when he officially introduced the iPhone, its music capabilities, and its brand-new touch screen. With the iPhone, you can take your music on the go—your whole CD collection if it'll fit—and it's all accessible with just a touch or a flick of your finger.

This chapter shows you the easiest ways to transfer your music stash to your iPhone, shows you how to navigate your music and playlists once they're on your phone, and offers some tips on keeping everything organized.

Moving Your Music

Apple has made it easier than ever to work with
your music files. Getting music into your computer? Just a click of a button. Transferring your
favorite music to your iPhone? That's just as easy.

Importing Music from CDs

Importing music from CDs into iTunes is a snap.
First, be sure your computer is connected to the
Internet; this allows iTunes to automatically list the
track information for nearly any CD you put in so
you don't have to type in the information yourself.

Incoming Call

If you can't get online when you're
importing a CD, you can also get the
track names later. Just highlight the track
names you need, go to the Advanced
menu at the top of the iTunes interface,
and click Get CD Track Names, all while
connected to the Internet.

Put the CD in your computer drive and, after a
moment, iTunes acknowledges the CD and looks up
the track information for you. The CD with the full
track listing appears in iTunes. In the lower-right corner, an Import CD button replaces the Browse icon.
Click on Import CD to copy your whole CD into
iTunes. If you only want to copy a specific song, click
on that song title and then click Import CD.

CD track list

import process bar

Import CD button

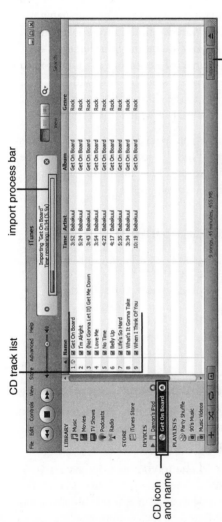

CD icon and name

Import your CDs into iTunes by clicking on the Import CD button.

It can take several minutes for your computer to import the music from the CD, depending on your CD drive and computer speed. The face panel shows the CD progress and the time remaining, and if you have your sound on, you'll hear a "ding!" when iTunes is finished importing the CD. You can also track the progress song by song with the green checkmark icons to the right of the song name. Now the music is part of your iTunes library.

Removing songs from your iTunes library is as easy as adding them. Find the song(s), click on the title, and hit the Delete key. This completely removes the song from your computer.

Music to Your Ears

You'll note in the preceding figure that the album art isn't showing. You can request iTunes to go find it for you by going to the Advanced menu and selecting Get Album Artwork, or you can find it yourself using the instructions in Chapter 3.

Adding Music to Your iPhone

Here's what you've been waiting for, right? It's time to load up your iPhone with as much music as it can hold. Here's how:

Connect your iPhone to your computer via a USB or FireWire cord. An icon shows up on the left in the Devices section. If this is your first time

connecting your iPhone, Apple will ask you to
register the device. Registering helps validate your
warranty.

Incoming Call

The following photos show the iPod
icon, but on your computer, it will be
the appropriate iPhone icon.

If any songs are already on your iPhone, a list of
them will appear in iTunes. At the bottom of the
screen you'll find information on the remaining
space available on your iPhone, as well as the number of songs and hours (or days!) of music space
available.

Transferring music to your iPhone is the same as
creating a playlist (remember that from Chapter
3?). Go to the Library and find music you'd like to
add. Click on the song title, drag the music into the
left column over the iPhone symbol, and release to
add it to your portable collection. It also transfers
the album art, if any, to your iPhone so you can
see it when you play the song or flip through your
library with Cover Flow.

To add a whole playlist to your iPhone, just highlight the playlist name on the left side and click and
drag it onto your iPhone icon.

Delete music by clicking on your iPhone, finding
the song(s) you want to remove, and hitting the
Delete key. This only removes the song from your
iPhone, not your computer.

iPhone/iPod icon

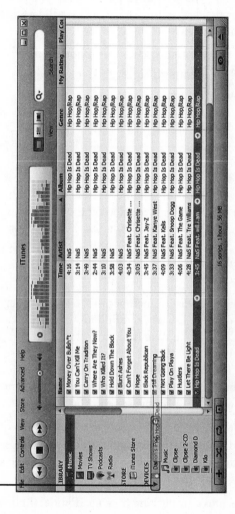

Highlight and drag your favorite songs onto your iPhone via iTunes.

Playing Your Tunes

To play music on your iPhone:

1. Turn on and unlock your iPhone.
2. Press the iPod icon at the bottom right of your home screen.

You'll find a text listing of all the artists loaded onto your iPhone. Flick your finger up to scroll down the list. Flick your finger down to scroll up the list. See the alphabet on the right side? Tap a letter and a listing of the artists saved under that letter comes up. Double-tap the artist to pull up his or her music.

At the bottom of the screen, you'll see five icons:

iPod Icons	
Playlists	Your iTunes playlists
Artists	A list of artists; the default display
Songs	List by songs instead of artists
Videos	Watch your iTunes videos
More	Additional options

If you don't like the listing by artist, tap the Songs icon and the iPhone gives you a list by song title. (See Chapter 6 to learn more about watching videos.)

When you decide on a song, the iPhone begins
playing it for you. Holding your iPhone vertically
gives you a display similar to the regular iPod, but
turn it vertically and it goes into a Cover Flow display similar to iTunes.

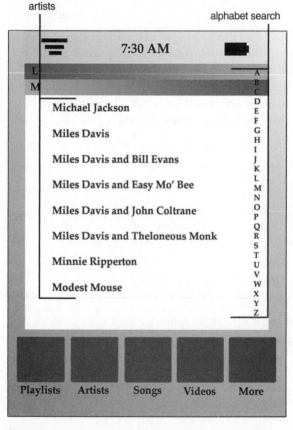

Tap a letter to find an artist alphabetically.

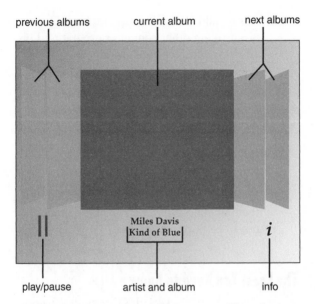

previous albums   current album   next albums

Miles Davis
Kind of Blue

i

play/pause   artist and album   info

Flipping through your music collection is easy. All it takes is a flick of your finger.

To listen to your music collection, press the play/pause icon in the lower-left corner to stop or start your music. Touch the *i* in the lower-right corner to get more information on the music you're hearing.

Organizing Your Music Collection

With music in iTunes, some on your iPhone, and some on your iPod, not to mention CDs and play-lists here and there, you can easily get musically unorganized. Here are some tips to keep your music collection in order:

◆ Always add your CD music through iTunes, not through other computer programs. This allows for cleaner organization, as all your music is saved in the same area on your computer.

◆ Delete your music through iTunes, not directly from your hard drive. If not done through the music program, iTunes still lists the now-unplayable track and you have to delete it in iTunes anyway.

◆ Find track names through iTunes. Typing them in yourself is an error-prone practice, not to mention a tiring one.

The Least You Need to Know

◆ The iPhone works with iTunes in the same user-friendly way the iPod does.

◆ iTunes automatically downloads song information and art for the CDs you import.

◆ Getting music into your iPhone from iTunes is easy—just drag and drop.

◆ Navigating and playing your music on your iPhone is as simple as pointing your finger.

◆ Keeping your music organized helps you better enjoy your music.

The iTunes Store

In This Chapter

- ◆ Downloading music, videos, and other media with iTunes
- ◆ Limitations of music use
- ◆ Cool new iTunes features

As you learned in earlier chapters, you use iTunes to import music from a CD into your computer and then transfer it to your iPhone. With the iTunes Store, you can browse for, purchase, and download music, videos, podcasts, and more—all ready to go onto your iPhone.

This chapter shows you how to download multimedia from the store, tells you what you need to know about digital music and videos, and explains the limitations of use.

Downloading from the Internet—
Legally

In 2003, Apple started the iTunes Store (then called the Apple Music Store), the first successful paid online music service. In a little over 2 weeks, more than 2 million songs were purchased, each for less than $1 a pop. On February 25, 2006, the iTunes Store sold its *1 billionth* song.

Besides convenience, the popularity of the iTunes Store can be attributed to a serious crackdown on illegal online music. The legality of using file share programs to get music was questionable in the 1990s, but around 2002, the Recording Industry Association of America (RIAA) began suing down-loaders several thousand dollars *for each illegal song obtained*. The iTunes Store provides an easy, afford-able way to get digital music.

The best part about the iTunes Store is that you can buy whole albums, usually for $9.99, or just a single song, for 99¢. This eliminates the need to buy an entire CD just because you like one or two songs.

> ♪ **Music to Your Ears** _____
>
>   If you initially purchase one or two
>   songs, you can later purchase the
> rest of the album at a discount, thanks to
> Complete My Album. Check out the "New
> Features" section at the end of this chapter
> for more information.

In 2005, Apple began offering music videos for download and later added TV shows and full-length movies, to either watch on your computer, video iPods, or your iPhone. Videos are priced at $1.99, while movies are around $9.99. TV show prices vary widely.

Shopping the iTunes Store

Shopping at and purchasing from the iTunes Store is, like most online shopping, easy—and like other Apple products, very user friendly. All you need are iTunes, an e-mail address, an active credit card, and an Internet connection.

Setting Up Your Account

To access the store, while connected to the Internet, click on the iTunes Store icon in the left column of the main iTunes screen. If this is your first visit to the store, iTunes will ask you if you'd like to set up an account. Just provide iTunes with an e-mail address, a password, and a credit card number. These preliminaries lead you to the iTunes Store front page.

Incoming Call

You have to set up an account to purchase music. However, after setting up your account, you can keep track of purchases and if you connect it to a credit card, you can buy music with one click.

Going Shopping

The iTunes Store front page updates regularly with new releases, top song download listings, exclusive iTunes-only tracks (songs not available on CD), and more. Navigate among music, movies, TV shows, etc. via the iTunes Store box in the upper-left corner of the front page. The front page also features audiobooks, music videos, and other multimedia for download.

For example, clicking on Music brings you to the music home page. Here you'll see New Releases, Top Songs, Top Albums, What's Hot, and more. You can browse by genre by scrolling through the Genres box on the left side of the page.

Crossed Signals

As you're browsing music, you'll see that some genres such as hip-hop and rock offer Clean and Explicit versions of singles and albums. (A small tag appears next to the album art.) Be sure you know what you're getting!

You can also search the store using the Search bar in the upper-right corner. This works just like searches in your iTunes library (see Chapter 3). Just type in your favorite artist or song, and iTunes lists the top albums, songs, and artists related to your criterion at the top, with every related song on the bottom half of the screen. Each song lists its title, playing time, artist, album, relevance, and song price. Double-clicking on a song or highlighting it and pressing play gives you a 30-second preview snippet.

account information

new releases

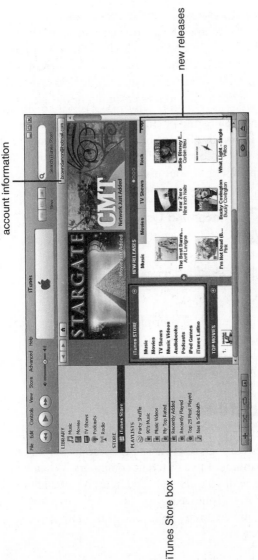

iTunes Store box

The iTunes Store front page shows the hottest music available for download, along with other multimedia.

Use the back arrow in the upper-left corner of the iTunes Store to go to the previous query and the nearby Home button to go back to the front page. The other side of the screen shows the current account, e-mail address, and, if a gift certificate has been entered, the money remaining in the account.

Pressing Buy Song or Buy Album downloads the item into your iTunes library. The cost is subtracted from your account or billed to your credit card. A receipt is sent to your e-mail account within a week.

Incoming Call

Search for movies, videos, audio books, and podcasts the same way as instructed for music.

Now you can drag your purchased item(s) to your iPhone when you want to listen or watch on the road. (See earlier chapters if you need a refresher on how to do this.)

Limits of Use

To thwart piracy, songs, videos, and podcasts purchased through the iTunes Store can be played on a maximum of five different computers. When you play one of your downloaded items on a buddy's computer, iTunes asks you if you'd like to allow his or her computer to be one of the authorized computers.

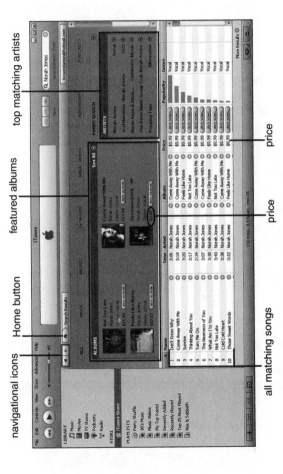

You can find songs in several different ways at the iTunes Store.

However, if you'd just like to *play* the item on your friend's computer, hook up your iPhone to his or her computer, find it in iTunes, and play it. The song, video, or podcast is automatically removed after you disconnect your iPhone.

> **Music to Your Ears**
>
> All your purchases from the iTunes Store are listed under the Purchased Music and Purchased Videos icons in your library.

Gifts and Allowances

Gift cards or certificates are easy to give, easy to receive, and easy to spend. And like most other retailers, the iTunes Store lets you use these pre-paid cards or send gift certificates and allowances through iTunes.

Redeeming Gift Cards

iTunes gift cards are often one of the top-selling gift cards of the Christmas holiday season. Here's how to redeem a card:

1. Click on the iTunes Store icon in the left column.

2. In the top-right Quick Links box, click on Redeem.

3. Type in the 16-digit code on the back of your prepaid card.

Your account is credited with the amount on the
gift card, and you can go shopping. Apple deducts
from the gift certificate as you make music purchases. If you try to make a purchase more than
what's remaining on the gift certificate, Apple
deducts the remaining balance from the credit card
on file for your account.

Incoming Call _____

In 2006, Apple added a Gift This
Music icon next to albums and videos.
Use this to buy a specific item for someone
instead of just giving a gift certificate.

Buying and Redeeming Gift Certificates

Gift certificates are the same as prepaid cards, but
they don't have to be physical. You can choose to
have a certificate sent to the gift recipient, but you
can also e-mail it to them.

To send a gift certificate:

1. Click on the iTunes Store icon in the left
   column.

2. In the top-right Quick Links box, click on
   Buy iTunes Gifts.

3. The next screen gives you a few options:
   iTunes Gift Cards; Printable Gift
   Certificates; Email Gift Certificates; Give
   Specific Music, TV Shows, and Music; or
   Allowances.

Crossed Signals

Apple sends the gift certificate as soon as you purchase it, so be sure to wait for the day you want the person to receive the gift.

If you choose Email Gift Certificates, Apple asks for your name, the recipient's name, and his or her e-mail. You can choose an amount between $10 and $200 and include a personal message. Your credit card is charged, and iTunes sends the gift certificate right away.

If you decide to print the gift certificate, Apple asks for your name, the recipient's name, and the gift certificate amount, from $10 to $200. You can also include a two-line personal message. Apple then gives you an electronic document you can print yourself.

If you decide to mail the gift certificate using the United States Postal Service, Apple connects you to its special Internet page. From there, you give your name and the recipient's name, the gift certificate amount from $10 to $200, and a personal message. Then you give the recipient's full address for mailing.

You redeem gift certificates just as you do gift cards.

Creating an Allowance

The iTunes Store lets you create an allowance so others can purchase iTunes music without having to have your credit card information at hand.

To set up an allowance:

1. Click on the iTunes Store icon in the left column.
2. In the top-right Quick Links box, click on Buy iTunes Gifts.
3. At the bottom of the next page, click on Set up an allowance now.
4. Type in your name, the recipient's name, the monthly allowance amount between $10 and $200.
5. Decide if they should receive the first installment now or on the first of the next month.
6. Type in the recipient's Apple ID or, if they don't have one, create one for them.
7. Type in a personal message for them if you like.

Apple deducts your credit card when a new allowance is given at the first of every following month. You can give the allowance to the person right away if you wish, but if you're making arrangements after the twentieth of the month, Apple automatically waits until the first of the next month.

Music to Your Ears

Creating an allowance is a great way to restrict how much money your kids spend on music. You can add a certain amount for them to use, and when it's gone, it's gone—with no surprises on your credit card bill!

New Features

Apple is constantly updating the iTunes Store to make it a more pleasurable experience. Here are some new features you might have missed.

MiniStore

While running iTunes and connected to the Internet, you might get a notice at the bottom of your jukebox asking if you want to activate the MiniStore. The MiniStore is like your own personal shopper. Play a certain song, and the iTunes Store recommends other, similar titles you might enjoy.

For instance, let's say you play Frank Sinatra's "New York, New York." If the MiniStore is activated, iTunes accesses the Internet and gives you a brief synopsis of the album you're currently playing and recommends four other albums for purchase. All the albums include their respective five-star ratings based on other purchasers' feedback.

Music to Your Ears

The MiniStore is a great way to learn more about some person, group, or music, but it's also terribly tempting to keep buying stuff—especially if you really love an artist or a genre!

When you get the MiniStore notice, click on Turn on MiniStore to start it up. Otherwise, click Not Now and the notice goes away for a while. If you'd rather not be disturbed while listening, you can look up the album yourself later in the iTunes Store and find recommendations there, too.

Complete My Album

It's frustrating to buy a music single or two and, later realize that you really want the whole album. In the past, you had no reprieve: you just had to pay the whole price for the album you wanted. But not anymore, thanks to Apple's Complete My Album feature, which allows you to buy the rest of the album at a discount.

For example, imagine you bought the previously mentioned Frank Sinatra song "New York, New York" as a single from the album *Frank Sinatra's Greatest Hits*. Let's say the single was 99¢, but the album is $9.99. The next time you go to the iTunes Store and look at *Frank Sinatra's Greatest Hits*, it won't be listed as $9.99. Instead, you'll see the "Complete Your Album for $9.00"—the price of the album minus what you already paid for one of the singles—followed by the usual Buy Now button.

Watch out, though: Apple only gives you 180 days, or about 6 months, to buy the album at a dis-counted price. After that, you're back to full cost.

Crossed Signals

Sometimes you'll find multiple versions of the same album. If you're purchasing the full album of a previously bought single, be sure you're purchasing the right album to get the discount.

The Least You Need to Know

◆ You need iTunes, e-mail, Internet access, and a credit card to use the iTunes Store.

◆ Singles cost 99¢, albums run $9.99, music videos are $1.99, and TV shows vary in price.

◆ You can play purchased music and videos only on five different computers.

◆ Audiobooks, music videos, movies, and other multimedia are available at the iTunes Store.

◆ The Complete My Album option lets you download the rest of the album at a discount if you've already purchased a single from it.

6

Photos and Videos

In This Chapter

- ◆ Loading your iPhone with movies and videos
- ◆ Watching movies and videos on your iPhone
- ◆ Taking and managing photos on your iPhone

Sure, many cell phones come standard with a camera these days, the iPhone included. But how many other cell phones or even smart phones have the capability to play videos and full-length movies—and boast a 3.5-inch widescreen to watch them on? Let's see, there's the iPhone and ... well, the iPhone.

Plus, with the iPhone's revolutionary touch screen, you can flip through your favorite movies and photos—literally—with a flick of your finger. Turn the page to learn how.

Loading Up

Apple makes it easy to load up your iPhone with videos. In fact, the process is very similar to how you transfer your favorite music.

Connect your iPhone to the computer using a USB or FireWire cord. Give it a second for an iPhone icon to pop up in the Devices section. Apple asks you to register the iPhone if this is your first time connecting it to a computer.

iTunes shows you how much memory (space) you have on your iPhone as well as the amount of space available. Songs take a few megabytes, but music videos and feature-length movies take several hundred.

Crossed Signals

Feature-length movies can take hundreds of megabytes and fill up your iPhone quickly, which means less room for songs and TV shows. Try to keep longer videos on your phone only as long as you're watching them. They're still available on your computer's iTunes for when you want them later.

To transfer videos and movies, go to the Library and find the stuff you want to move to your iPhone. Click on the program title and drag it into the left column over the iPhone symbol. Let go of the mouse button and that's all it takes to transfer it to your iPhone. The appropriate cover art transfers, too.

To take a movie or video off your iPhone, simply find the file and hit Delete. As with music, this only removes the file from your iPhone, not your computer's iTunes collection.

Playing Videos and Movies

The iPhone plays videos and movies just like the video iPod does. And with the touch controls, navigating through your lists of movies and videos is quick and easy.

> **Music to Your Ears** _____
>
> Music and movies were originally available only through the iTunes Store, but Apple is expected to make the iTunes Store available directly through the iPhone in the near future. You can download songs and movies without first hooking up to your computer. Cool, huh?

Your movies, like your music (see Chapter 4) are available under the iPhone's iPod function:

1. Turn on and unlock your iPhone.
2. Press the iPod icon at the bottom right of your Home screen.

3. Touch the video you want to watch.

4. If you haven't already, turn your iPhone landscape, or horizontally. This gives you the widescreen view to see more of the action.

The iPhone will spread your whole movie or video collection out like a flipbook, and you can simply use your index finger to "push" the various album covers or video titles along like a jukebox. When you find what you want, double-tap the cover, and then, if necessary, double-tap on the specific video or episode.

 Incoming Call

You can download movie covers or video art through iTunes. Check out Chapter 3 for more information.

As with most all other things Apple, on the iPhone's video interface, some icons help you watch or listen:

Video Playback Icons	
Time bar	Shows the remaining time
Double right arrow	Goes to the next song or episode
Double left arrow	Goes to the previous song or episode
Double vertical bars	Pauses the selection

movie cover or program art

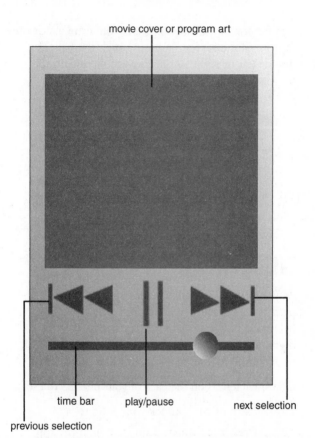

time bar   play/pause   next selection

previous selection

Apple has made it simple to control music and video playback on the iPhone.

You'll notice a ball moving along the time bar at the bottom of the screen. You can touch the ball and drag it left or right to fast forward or rewind the playback.

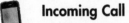

> **Incoming Call**
>
> Generally, flipping the iPhone horizontally into widescreen mode is better for movies. You can flip the iPhone between horizontal and vertical anytime, even when you're in the middle of listening or watching.

Taking Pictures

As you know, the iPhone comes with a 2 MP (megapixel) camera, which, admittedly, is a few notches below the now-average 3 MP resolution of most digital cameras. But it's still plenty powerful for a cell phone camera. And it comes in a killer package!

The small lens is on the back of your phone. To use it:

1. Turn on and unlock your iPhone.
2. Press the Camera icon located at the top right of the Home screen.
3. You can then take pictures with the phone, using the touch screen to aim and shoot (instead of the traditional viewfinder hole in most cameras).

Photos are automatically saved on your iPhone photo library, where you can flip through the pictures like you flip through your music and videos.

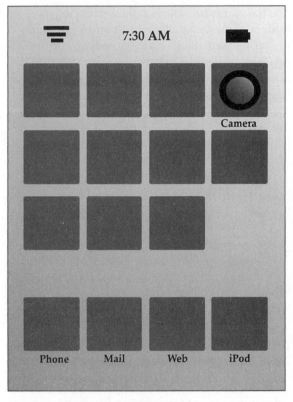

Tap the camera icon and start shooting!

To look at a particular shot, double-tap the picture. You can then examine it closer by "pinching" it—taking your thumb and your index finger on opposite ends of the picture and stretching it to the borders of the touch screen. To zoom out, take your thumb and index finger on opposite ends and push them closer to each other. Don't worry: the zooming and shrinking is temporary, so it won't permanently affect your picture.

 Incoming Call

> As you may have noticed now, nearly every item on your iPhone touch screen (except for the program icons) can be stretched, shrunk, and manipulated. Try it on your own, and see what you can play with!

And like your music and videos, turning the iPhone horizontally gives you a nice landscape view.

As of summer 2007, Apple was updating the appropriate iTunes software to accept iPhone photos. Visit the Apple website at www.apple.com for the latest info.

The Least You Need to Know

◆ You can flip through your video and movie collection like a jukebox with a flick of your index finger.

◆ Turn your iPhone horizontally for better movie viewing and faster video selection.

◆ Push and hold the ball on the time bar at the bottom of the playback screen to fast forward or rewind a video.

◆ Use your fingers to "pinch" your photos— that is, stretch or shrink them.

Chapter 7

Getting Online

In This Chapter

- ◆ The Internet in your pocket
- ◆ E-mail from wherever you are
- ◆ Instant messaging with the iPhone

Most cell phones these days are equipped to handle instant messaging, and a few have limited web browsers. But the iPhone blows them all out of the water—as usual. In this chapter, we go over getting on the Net; e-mailing your friends, family, and colleagues; and even talking to them "live" via iPhone's instant messenger.

Taking the Web on the Road

One of the coolest things about the iPhone is the capability to get online while you're out and about. With the iPhone, you can catch up on your favorite websites pretty much where you want and when you want.

But here's the cool thing: the websites you can access via your iPhone are the same websites you can access via your laptop. They're not baby versions. They're not text-only versions. They're not stripped-down versions. They're the *real* thing—the full-photo, full-color, real thing. Apple spared no expense with the iPhone's web browsing capabilities … but that shouldn't surprise anyone by now.

Crossed Signals

Like other cell phones, your phone carrier might charge you Internet time per minute. Be clear about what you'll be paying to avoid receiving a scary monthly bill!

Connecting Wirelessly

Your iPhone comes ready for Internet access and, like any wireless-enabled laptop, it automatically connects to any nearby Internet hubs. The iPhone uses 802.11b/g Wi-Fi—a big tech term that basically means it can use any standard wireless Internet connection. And if you move from one Wi-Fi network to another (say, if you're on the train or in a car—not driving!), the iPhone automatically searches for and joins the new network.

Blue What?

The iPhone also uses Bluetooth 2.0. Have you seen people wearing wireless *Star Trek*–looking earpieces— or have you seen or heard someone apparently talking to him- or herself and didn't catch the earpiece hidden under their hair (or, worse, thought they were talking to you and talked back)? Those wireless earpieces are Bluetooth devices.

Bluetooth is a wireless protocol, or standard, that enables two equipped devices to communicate over a short distance. In the case of the funky earpieces, the Bluetooth-enabled cell phone transmits the phone conversation to the Bluetooth-enabled ear device, allowing the user to talk on the phone without having to have it up against the ear or use any connecting wires.

Music to Your Ears

Apple has a sharp, less-obvious Bluetooth headset for the iPhone. It's sold separately.

Bluetooth is also a cool way to transfer other information between two devices. Apple is expected to take more advantage of the iPhone's Bluetooth compatibility in the near future to transfer contacts and other data.

More Tech Talk

Finally, the iPhone is a quad-band GSM phone, meaning it can be used not only in America, but in other countries, albeit in a limited way. Apple says it will be expanding its markets as time goes on.

Crossed Signals

Quad-band phones work best in Western European countries, but even there, the functionality may be limited to making phone calls and not the other cool iPhone Internet and e-mail options.

Browsing with Safari

The iPhone comes with Safari, a web browser popular on the Mac. And like other iPhone options, Apple went all out. The version of Safari on the iPhone is a full version of Safari, like you'd find on your computer, not a stripped-down version.

While other cell phones offer Internet access, most don't give you the same websites you experience on your home computer. Instead, major companies such as Yahoo! and Microsoft create special cell phone–friendly websites with less text and pictures. Not so with the iPhone and Safari. With this pair, you get the real deal. What you see on your laptop is what you see on your iPhone.

Here's how to hop online:

1. Turn on and unlock your iPhone.
2. Press the Web icon located at the bottom of the home screen.
3. Use Safari to surf wherever you want to go.

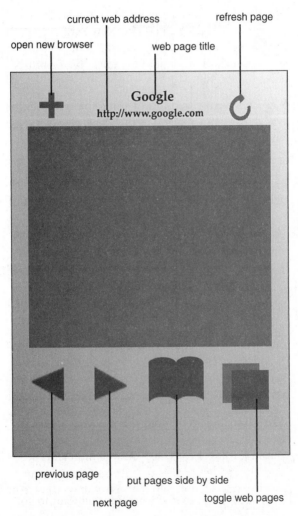

With Safari, you can surf the web on your iPhone the exact same way you do on your computer.

At the top of your screen you'll see the name of the current web page, such as The New York Times or Google. Directly below the title is the website address, such as http://www.nytimes.com or http://www.google.com.

You can navigate websites by using the various icons available. Here's a listing of the icons, clockwise from the top of your screen:

Safari Web Browser Icons	
Plus	Opens another browser window
Clockwise arrow	Refreshes the current page
Two layered squares	Looks at another active browser window (if available)
Book	Puts two pages side by side (if available)
Right arrow	Goes to next page (if available)
Left arrow	Goes to previous page (if available)

When using Safari on your iPhone, your finger is the same as a mouse. You simply double-tap with your index finger to "click" on web links and use the right and left arrows to traverse through your favorite sites. Double-tap the web address to type in a new location.

You can also stretch and shrink certain areas of the web screen by "pinching" them. To expand a particular section, put your thumb and index finger at the edges of that area and widen them until you reach the edge of the iPhone screen. To shrink things back to normal size, put your finger and thumb at the edge of the screen and pull them toward each other until the item is the size you want.

You can also do this with photos. See a particular web picture you like? Double-tap on the pic and the iPhone zooms it for you. Double-tap it again to go back to normal web viewing. The same can be done with small text.

The web browser works horizontally or vertically. Just turn your iPhone, and the view adjusts.

Crossed Signals

The iPhone browser won't work if you're not near a wireless Internet connection, just like if you were using a laptop.

Google Maps

In 2005, Google tested out a new feature of the Google empire called Google Maps. It quickly caught on as a favorite way to find maps, store and business locations, and driving directions. And thanks to Apple's partnership with Google, not only is Google on all iPhones, so is Google Maps.

And like Safari, this isn't a stripped-down version.
Google Maps on the iPhone is a virtual carbon copy
of the Google Maps you get on your computer.

> **Music to Your Ears** _____
>
> Test-drive the original Google Maps at
> maps.google.com.

Here's how easy it is to use Google Maps:

1. Turn on and unlock your iPhone.

2. Press the Maps icon on the home screen.

3. When Google Maps has loaded, type in your
   desired location by double-tapping the location bar at the top of the touch screen. You
   can type a specific destination, such as "1115
   E. Mulberry, Chicago, IL" or something
   more general like "Chicago." The location
   will appear onscreen.

It's just as easy to get to-from directions:

1. Double-tap the location bar in Google Maps.

2. Type your current location in "From:".

3. Type your destination in "To:". The directions will pop up.

You can also find specific locations on Google Maps
by typing in an area and type of destination, such
as "Sushi Restaurants in Los Angeles." A group of
locations will pop up on the map. Press a specific
location to get information on it.

Google Maps also has a satellite option that shows a real satellite picture of the area you're exploring. It's usually very detailed! Touch the Satellite tab at the bottom to change the view.

Crossed Signals

Google Maps gives driving directions, not for walking or public transportation.

Staying in Contact

But wait, there's more! You can also use the iPhone to stay in touch with friends, family, and colleagues by using e-mail or instant messenger—two mediums that, until the iPhone, have mostly been limited to home computers and laptops.

iPhone E-Mail

The iPhone uses something called rich HTML e-mail, a system that acts as a courier for your current e-mail account, fetches your e-mail for you, and delivers it to your iPhone. You just need to type in your e-mail address, password, and other appropriate information to get the service set up.

Music to Your Ears

Rich HTML is better than plain e-mail because you can use special text features like italics, various text sizes, and limited pictures and graphics within your message.

iPhone works with most popular e-mail systems, such as Yahoo!, AOL, and Google's Gmail—or to get techie, any systems that use a *POP3* or IMAP format. Check your iPhone manual or www.apple.com to find out if your e-mail is supported.

> **iTerms** _____
>
> **POP3** stands for Post Office Protocol, a standard established to send e-mails. POP3 is the third, and most popular, generation of the standard.

To check your e-mail through the iPhone:

1. Turn on and unlock your iPhone.
2. Press the Mail icon located at the bottom of the home screen.

> **Music to Your Ears** _____
>
> Both Yahoo! and Google offer free web-based e-mail services that work with your iPhone. Check out www.yahoo.com or www.gmail.com for more information.

Instant Messenger

AOL, Yahoo!, and other major e-mail systems support instant messaging (IMing), a way to text notes back and forth with someone. It's on the computer, similar to e-mail, but live—just like a real conversation. If you're one of the millions of users familiar with instant messaging, you'll feel at home with iPhone's *SMS*/instant messenger program.

iTerms

SMS, or short message service, is a term that describes any text messaging done between two cell phones.

To use instant messenger:

1. Turn on and unlock your iPhone.
2. Press the Text icon on your home screen.
3. Type in the phone number or use the address book to find the person you'd like to have a conversation with.
4. If they're available, the conversation will begin.

The top half of the screen shows the conversation. On the bottom half of the screen the keyboard pops up. (The forward-thinking folks at Apple made it so your phone knows when you need to type and provides the keyboard, and when you're done, it goes away.) Your friend's messages are posted to the left, while your comments are to the right. Use the keyboard to type in your message and then press the Send button.

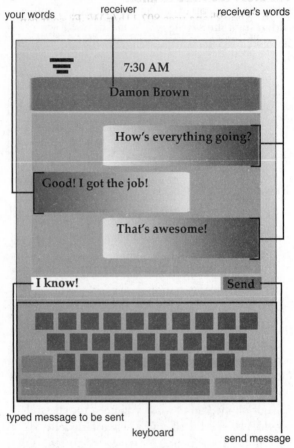

your words · receiver · receiver's words

7:30 AM

Damon Brown

How's everything going?

Good! I got the job!

That's awesome!

I know! Send

typed message to be sent · keyboard · send message

Have a live conversation with a friend with the iPhone's real-time text-messaging capabilities.

The Least You Need to Know

- ◆ The iPhone uses 802.11b/g Wi-Fi, so it can connect to virtually any wireless hotspot available to a laptop.

- ◆ Yahoo!, AOL, Gmail, and most other e-mail services are compatible with the iPhone.

- ◆ Google Maps on the iPhone works pretty much the same way it does online at maps. google.com.

- ◆ iPhone's instant messenger lets you talk with another person (or people) via real-time text messaging.

iPhone Accessories

In This Chapter

- What's included with your iPhone
- Speakers, Bluetooth headsets, and other iPhone add-ons
- Other fun iPhone accessories

Like the iPod, the iPhone is one of the most customizable devices out there. It comes packed with everything you need, but Apple and other companies offer dozens of accessories that help you tailor your iPhone to your needs. In this chapter, you learn how to maximize your iPhone experience.

What Your iPhone Comes With

Your iPhone comes with the basic necessities you need:

- iPhone unit
- Ear bud headphones
- USB cord
- Wall plug

That's enough to get you started and allow you to connect your iPhone to your computer, get music and videos on your iPhone, recharge your iPhone when necessary, and listen to what's playing. But there's so much more fun stuff you can get for your iPhone.

Music to Your Ears _____

Check out Apple.com for the latest iPhone accessories. A quick Google search for "iPhone accessories" should yield some fun results, too.

Stuff to Buy

Many products were available for the iPhone before it was even released. As soon as the dimensions of the phone were announced, manufacturers started cranking out cases, covers, and other iPhone add-ons.

The following sections cover some of the cool items you can get to personalize your iPhone. Check out Appendix B for the websites of some iPhone-accessory manufacturers.

Covers and Cases

Apple and several third-party companies have various kinds and types of sleeves that protect your iPhone from wear and tear. Cases add some bulk to your tiny iPhone, but with the sensitive touch screen, you'll need something to protect it.

Headphones

Like the iPod, the basic iPhone headphones are ear buds—tiny round plugs that stick in your ear. If you aren't a fan of ear buds, most any another basic set of headphones will work in your iPhone. You can use the same headphones for your iPhone that you use for an iPod or other portable music devices like tape and CD players.

Music to Your Ears

Apple is planning ear buds for the iPhone with a built-in mic so if you get a call while you're listening to music, you can simply talk without removing your headphones and the mic will pick up your voice. You'll hear your caller through your ear buds. These might become the standard headphones for the iPhone.

Bluetooth Headsets

Apple and Gomadic are among the many companies offering *Bluetooth* headsets. Reminiscent of something you'd see on *Star Trek*, these devices fit securely into your ear and allow you to have conversations completely wirelessly. You can leave your iPhone in your pocket and still talk while having both hands free to shop, work out, drive, etc.

The Bluetooth headsets do run on batteries, so like your iPhone, you have to recharge them. And they can be on the pricy side sometimes.

> **iTerms** _____
>
> **Bluetooth** is a wireless way computers can communicate with each other.

Speakers

Bose, inMotion, and other companies have special speakers especially made for the iPod, some of which work with the iPhone, too.

You have two solid bets: Apple and other companies have a setup that uses wireless signals to connect your iPhone/iPod to speakers around the house. Also, traditional portable music device speakers that just plug into the headphone jack work well with the iPhone.

> **Crossed Signals** _____
>
> If you're looking to buy speakers for your iPhone, avoid the ones that are structured with a space in the middle to plug in your iPod securely—the iPhone might not fit.

Travel Accessories

With the different kinds of power outlets in various countries, one of the most useful accessories for your iPhone is the Apple World Travel Adapter Kit, a smattering of outlet plugs that allow you to recharge from nearly anywhere on the planet.

To recharge your iPhone while you drive, invest in some car recharging accessories, such as the Belkin Auto Charger, that fit in your cigarette lighter.

You also might want to pick up some car accessories, such as the Sony Car Cassette Adapter, which allows you to listen to your iPhone through a tape deck. Devices such as the iTrip can even let you play music wirelessly through your car radio if you don't have a tape deck.

The Monster iCase Travel Pack is a handy case that keeps all your accessories organized.

Apple TV

Apple TV is a device that enables you to play movies and videos from iTunes, your iPod, or your iPhone on your living room television set. This is a really cool way to play movies you might only have stored on your iPhone or in iTunes on a screen larger than your 3.5-inch iPhone screen or your computer monitor.

Other Fun Stuff

There's no shortage of fun (and funny) accessories out there for your iPhone. For instance, the Cellulounger is a cute miniature lounge chair to rest your device in when not in use. A fun item from Apple is the iPod Sock, a knitted cover that keeps your iPod snug. Although it's designed for the iPod, it works with the iPhone, too.

The Least You Need to Know

◆ Your iPhone comes with the unit, head-phones, USB cord, and wall plug.

◆ Covers are available to protect your iPhone.

◆ Travel accessories allow you to wirelessly play your iPhone music through your car stereo.

◆ Apple TV can connect your music and videos wirelessly to your TV set.

Chapter 9

Caring for Your iPhone

In This Chapter

- ◆ iPhone maintenance tips
- ◆ When it's okay to use third-party equipment ... and when it's not
- ◆ What do I do when ...?

Taking good care of your iPhone is very important, especially with its high-tech insides carrying your music, photos, videos, contacts, podcasts, etc. In this chapter, I give you tips on keeping your iPhone in good working condition, as well as help troubleshoot some problems you might encounter with your new phone.

Care and Feeding of Your iPhone

If you already have a cell phone, you probably understand how banged up and bruised the device can get over time. Now imagine that phone having a sensitive screen and $500 or $600 worth of hardware inside. Ouch. Fortunately, with some care, you can help prolong the life of your iPhone.

In That Case ...

To avoid scratches and cracks on the all-important touch screen, get some type of case for your iPhone—immediately. Months before the iPhone hit the market, accessory manufacturers started churning out iPhone cases, so you're sure to be able to find several either online or in stores.

Keep It Cool

Apple recommends running your iPhone only when it's between 32 and 95 degrees Fahrenheit (0 to 35 degrees Celsius), which may not always be possible in your climate, but still something to keep in mind. When it's not being used, keep your iPhone in a place between –4 and 113 degrees Fahrenheit (–20 to 45 degrees Celsius)—a much easier range to follow. Remember this if you decide to take your iPod to Siberia or the Sahara.

Direct sunlight can discolor (not to mention bake) your iPhone, so be careful where you leave it, especially in your car. The same goes for your accessories. Your power adapter is particularly sensitive.

Keep It Clean

Even with careful handling and storage, your iPhone will probably get gunky from time to time and need to be cleaned. Use a slightly damp, soft cloth to clean your phone, preferably a cloth that's lint- and static-free. Paper towels and facial tissues might leave lint or, worse yet, scratch the touch screen.

Crossed Signals

It might be common sense, but it's worth saying: your iPhone is an electronic device, so keep it from getting wet at all costs.

Be sure to disconnect your phone from your computer or the wall outlet before you start cleaning it. And Apple suggests avoiding aerosol sprays, solvents, alcohol, and abrasives. They aren't really necessary. The cloth is sufficient.

Apple Power Cord: There's No Substitute

Non-Apple headphones, or those made by other manufacturers, are fine, but if you happen to misplace your iPhone power cord, especially the wall plug, do not go to the corner electronics store and get a replacement unless the store is an authorized Apple dealer and carries Apple products. Another brand wall plug could fry your iPhone. It's best to go to an Apple store or visit www.apple.com to purchase replacement parts.

Troubleshooting

The iPhone has its quirks just like any other computer device. That's to be expected, so don't panic if you run into some problems. The following pages offer some troubleshooting suggestions.

No Response from the Computer

If your computer doesn't respond when you hook up your iPhone, check that you installed the iTunes program from the included CD.

If you're all installed, check that the power cord wire is secure to the bottom of your iPhone. Remember, you need to press the small tabs on the sides of the cord's flat end to put it in the iPhone and then release the tabs to ensure a secure connection. Your computer will let you know either with a "ding!" or a visual indication when a connection is made.

Incoming Call

If you hear the "ding!" but see no change on your computer indicating it recognizes your iPhone, give it a second. It can take time for the computer to bring up iTunes and the iPhone information.

Worst-case scenario: turn off your computer, turn off your iPhone, turn on your computer, turn on your iPhone, and reconnect your iPhone.

No Sound Coming from iPhone

If you're using your headphones, double-check that they're pushed all the way in. Also confirm that the volume is high enough. Or give it a second;

the iPhone has occasional "hiccups" that pause it between songs, which may be why things got quiet.

Worst-case scenario: restart the iPhone.

The Least You Need to Know

- ◆ Keep your iPhone away from extreme hot and cold temperatures.
- ◆ Your iPhone *does not* like water. Keep it dry.
- ◆ Except for headphones, generally avoid third-party replacement parts and stick with Apple.
- ◆ If your iPhone is acting strange, disconnect (or reconnect) it to your computer.

Glossary

Don't know your SMS from your instant messaging? Think an iPhone carrier has a contagious disease? Technology can be hard to understand. Definitions are here to help.

AAC Advanced Audio Coding. This is the music format similar to the popular MP3 music format but is more advanced and generally has a higher sound quality. The iPhone and iPod are two of the few portable music players that use the AAC format, so users of other portable players are better off using MP3 format.

Apple Touch Wheel The iPod's control device, a sensitive, flat joystick. The cardinal directions correspond to Menu, Next/Fast Forward, Play/Pause, and Previous/Rewind, and the Select button is in the center of the wheel. Use your thumb to motion along the wheel clockwise or counterclockwise to navigate the menus.

Apple TV A product that lets you watch your iTunes visual media, like movies and music videos, on your television.

audiobooks The digital equivalent of books on tape, read by actors or the authors themselves. The iTunes Store has a large selection of recent books as well as classics you can download. Most books have large files and take multiple hours to listen to.

autofill The process in which iTunes automatically loads your iPhone or iPod with random music until it's full. You can have iTunes autofill from your entire library or from specific playlists.

AV cable Audio-visual cable. You can purchase the optional iPod AV cable to connect your iPod to a TV or projector for slideshows (or purchase Apple TV for a smoother setup).

backlight Lighting that illuminates your iPod display. Using the backlight continuously can cut your battery charge in half.

Backlight Timer A setting that shuts down your iPod backlight after a predetermined time to conserve energy. Change the Backlight Timer parameters under the iPod Settings menu.

Bluetooth A wireless way computers can communicate with each other. Like other cell phones, the iPhone can use the popular Bluetooth headset that fits in your ear to allow hands- and wire-free calling.

carrier Also known as a *phone carrier,* this is the company providing your cellular service. As of summer 2007, Cingular is the only carrier that supports the iPhone.

CD-R (Compact Disc–Recordable) and CD-RW (Compact Disc–Rewritable)   CD-R and CD-RW are two formats of recordable compact discs. CD-Rs can only be recorded on once, while CD-RWs can be erased and rewritten over again and again.

contacts   Your collection of phone numbers, which can also include home and work addresses as well as e-mail. If you're using a Mac, the iPhone transfers your contacts automatically.

crossfade   A disc jockey term for blending two records together during a music set. Crossfade blends the current song and the next song into a seamless mix. You can set the crossfade blending time up to 12 seconds.

downsize   To make a particular program window smaller. To downsize a screen, aim for the lower-right corner, hold down the mouse, and adjust the window to the size you like.

FireWire cord   An alternative to the USB cord to connect your iPhone or iPod to your computer. Most computers can use both USB and FireWire cords, but some only accept FireWire connections.

gigabyte (GB)   A unit of measurement for computer space. 1 GB equals about 1,000 megabytes (MB). On the iPhone, 1 GB holds about 250 songs.

hard drive   The main memory area where a computer stores information. On your iPhone or iPod, this is where your music, pictures, movies, audiobooks, etc. are stored.

headset jack The small hole atop the iPhone and iPod where you plug in your headphones. You can also plug in alternative devices, such as compatible speakers.

Hold switch A locking, or holding, option on the iPod that prevents any button interference. For instance, if you want to listen to music without accidentally pressing a button, push the Hold tab, located on the top of the iPod next to the headset jack, to the Hold position.

Home button Get lost in the iPhone menus? The Home button, located right below the screen when the iPhone is held vertically, brings you to the main menu.

instant messenger Software that lets you have a "live" text-only conversation with another person through your computer or cell phone.

iSync A Mac program that automatically synchro-nizes your different Apple software and products. Use it to coordinate information (music, photos, contacts, etc.) between your iPhone or iPod and your Mac. It's not available for the PC.

iTunes A music jukebox for your computer and the main interface between you and your iPhone and iPod. Through iTunes, you can recharge your iPhone and iPod, change preferences, and access the iTunes Store.

iTunes Store   The online store where you can
buy music (whole albums or individual tracks),
music videos, TV shows, movies, podcasts, and
audiobooks directly from Apple. To go to the store,
you must have iTunes installed, an Internet connection, and a major credit card or iTunes Store gift
card.

library   Your entire music collection on iTunes.
Libraries are broken down into playlists, specific
sets of songs you assemble yourself. Library can
also mean a group of podcast programs.

megabyte (MB)—A unit of measurement for computer space. A little over 1,000 MB make 1 gigabyte
(GB). On the iPhone, 512 MB holds about 120
songs.

MP3   A popular music format. The name stands
for Motion Picture Experts Group Audio Layer 3,
named after the coalition that created it. iPhone,
iPod, and almost every other portable music player
accept the MP3 format. However, the iPhone and
iPod use the AAC music format by default, a format not compatible with most other portable music
players. Go to the Advanced screen to change the
format.

On-The-Go Playlist   An impromptu playlist
you create on your iPhone or iPod, as opposed to
a traditional playlist you create within the iTunes
program. Add songs by finding them in your library
and holding the Select button until the title flashes.
Your iPhone or iPod saves the selected songs in
playlists called "On-The-Go 1" and so on.

Party Shuffle   A continuous, random song mix iTunes shuffles from your general library or a specific playlist. The party doesn't stop until you intervene.

playlist   A specific set of songs you put together. Playlists allow you to collect songs from your library that fit a certain theme, like dancing, and listen to them together. iTunes can also create playlists for you, called Smart Playlists, after you give it the parameters.

podcasts   "Radio" programs you can download and listen to on your iPhone or iPod or in iTunes. These regular, prerecorded shows are available through the Internet, with new programs arriving every day.

POP3   Post Office Protocol, a standard established to send e-mails. POP3 is the third, and most popular, generation of the standard.

ppi   Pixels per inch. This measures how detailed a particular camera's photos are. The higher the pixels per inch, the more memory the picture requires.

reception   The clarity of your iPhone calls, which can vary based on your location. A small meter in the upper-left corner of the iPhone display shows your current reception power.

remote port   A small hole parallel to the headset jack used for the iPod remote. The optional remote is positioned on your headphone wire, allowing you to control the iPod without having to use the Apple Touch Wheel. It is not available on all iPods.

SIM card   Subscriber Identity Module card. Used by all cell phones, this tiny memory chip stores your phone numbers as well as any personal data.

sleep   The temporarily off mode iPhones and iPods switch themselves to for conserving energy when the device hasn't been used in a matter of minutes, unless the default settings are changed. Larger iPods can be forced into sleep mode by holding the Play/Pause button for 3 seconds. The iPhone can be put to sleep (and woken up) by holding the Sleep/Wake button at the top.

Smart Playlist   An iTunes-created playlist based on your preset criteria. For instance, you can tell iTunes to collect all the jazz songs featuring Ella Fitzgerald along with Mel Torme. Taking advantage of Smart Playlists is much faster than manually creating a playlist.

smartphone   Any phone with functions similar to a personal computer.

SMS   Short Message Service, or text messaging. This is sending a brief text note to another phone.

subscriptions   Regular updates of particular podcasts you've signed up to receive.

text messaging   Also known as SMS (Short Message Service), this is sending a brief text note to another phone.

thumbnail   A small version of a photo. The iPhone and iPod use thumbnails so you can quickly scan through a group of small photos instead of having to flip through each larger one.

touch screen A computer monitor that reacts based on when and where you touch it. The iPhone uses a touch screen.

USB cord The wire connecting your iPhone or iPod to your computer. The alternative to a USB cord is a FireWire cord.

vCard A popular file format for personal contact information. Used by Microsoft Outlook and other e-mail/organizational programs, it is the only contact information format the iPod and iPhone understand.

WAV WAV(e) form audio format, created by Microsoft and IBM. This music format option works well with PCs, but the file sizes are larger than MP3 or AAC, making it a less-desirable choice for your iPhone or iPod.

Wi-Fi Any Internet connection that allows users to connect wirelessly.

Important Websites

If you received an iPod for Christmas one year
and went to a store a few days later to find a case
or other accessory, you probably found the iPod-
related shelves nearly bare. The same might well be
true when you get a new iPhone. That's where the
Internet comes in handy. In this appendix, I share
some websites that help you find some fun accessories, software updates, and more for your new
iPhone.

Altec Lansing inMotion

www.alteclansing.com

Altec Lansing's inMotion product line features a
slew of iPod accessories. Most are surround-sound
speakers that double as docking stations for your
iPod, such as the barrel-shaped iM7.

Apple iPhone

www.apple.com/iphone

Check here for the latest info on the iPhone, directly from Apple.

Apple iPod

www.ipod.com

If you need information about your specific iPod, this is the place to start. The Apple website has a great breakdown of the different models and a page dedicated to the accessories made especially for your type of iPod.

iPod Software Updates

www.apple.com/ipod/download

Let iTunes tell you when a new update to your iPod software is available. If you happen to miss an update, or if you're not set up to check for updates automatically, visit and bookmark this site for the latest downloads.

Apple iTunes

www.itunes.com

This extensive website answers all your iTunes questions. Links to the newest music and movies online are offered here, too.

iTunes Software Download

www.apple.com/itunes/download

Log on here to download the latest version of iTunes for your Mac or PC. (Handy if you happen to misplace the iTunes CD that came with your iPhone.)

iTunes Jukebox

www.apple.com/itunes/jukebox

Check here for more information on iTunes Jukebox.

iTunes Playlists

www.apple.com/itunes/jukebox/playlists.html

Questions about playlists? Tune in here.

Apple iTunes Store

www.apple.com/itunes/store

Different from the iTunes website, the iTunes Store site provides information specifically related to purchasing music, movies, TV shows, audiobooks, podcasts, and games online. Here you'll also find exclusive content, like an iTunes-only TV show episode or a special remix of a song.

iTunes Store This Week

www.apple.com/itunes/weekly

A constant flow of new music, videos, and audio-books are available on the iTunes Store this week site. Check it out to see the latest goods.

Apple Products Store

http://store.apple.com

Here you'll find an extensive online store where you can buy all the latest Apple products for your iPhone. You can also purchase iPods and accessories here as well as Mac computers and related accessories and software.

Apple TV

www.apple.com/appletv

With Apple TV, you can watch videos and movies from your iTunes library on your regular TV. You can also listen to music from your library on your TV with this device, if you're so inclined. Learn more about it here.

Belkin

www.belkin.com/iPod

Next to Apple, Belkin has perhaps the largest collection of iPhone and iPod accessories. Among the products available are replacement cables and carrying cases.

Bose

www.bose.com

Bose has an entire line of iPhone/iPod-related accessories, including headphones.

Gomadic

www.gomadic.com

With its extensive, specific catalog, Gomadic provides accessories for the iPhone as well as virtually every iPod in existence—all dozen-plus generations.

Incipio Technologies

myincipio.com

Incipio Technologies carries several Apple accessories for the iPhone and iPod.

iPodder

www.ipodder.org

iPodder is perhaps the best program for podcasting—that is, downloading free "radio" shows to iTunes, your iPhone, or your iPod to listen to later. The shows are prerecorded but usually have a regular schedule. You can download iPodder as well as new podcast programs from this free software website.

Marware

www.marware.com

Marware has carrying cases and other accessories for all the iPods (sans the Shuffle).

iTunes Pull-Down Menus

iTunes, like most Apple products, is designed to be very user friendly and uses icons and other visual cues to tell you what's going on. However, at times, you need something spelled out in plain English. That's when the pull-down menus at the very top of the iTunes screen come in handy. In this appendix, I offer a breakdown of the menu options for iTunes 7 and their various functions, for both Macs and PCs where they differ.

iTunes (Mac)

Apple provides extensive help documents for iTunes. Some require an Internet connection.

About iTunes

Copyright information.

iTunes Hot Tips

This links you to an Apple website with pointers on how to make the best of your iTunes setup. You need Internet access to view it.

Preferences ...

This brings up the powerful Preferences menu, where you can control your iPod options, CD-burning techniques, and more.

Shop for iTunes Products

This option takes you to the iPod Accessories website, where you can purchase the latest iPod products. You need Internet access to enter. (See Chapter 8.)

Provide iTunes Feedback

Here you can communicate with Apple: about a CD you think should be available through the iTunes Store, about an error in the music listing, and other kinds of feedback. Apple does not reply in most cases, but the company says all feedback is noted. You need Internet access to talk with Apple.

Check for Updates ...

Touch base with Apple to check for recent iTunes software updates with this option. Apple usually sends you a notice when a new iTunes update is available, but this option is convenient if you happen to miss the message. Internet access is required.

Quit iTunes

This is how you shut down iTunes.

File

Here you find things related to importing and exporting music, playlists, and burn options.

New Playlist

Choosing this option is the same as pressing the Playlist icon; it creates a Basic Playlist for you. (See Chapter 3 for more information on playlists.)

New Playlist from Selection

This is the same as pressing the Playlist icon, too. If you highlight a specific group of songs, iTunes will create a Basic Playlist starting with these songs. (See Chapter 3.)

New Smart Playlist ...

With this, iTunes creates a Smart Playlist. With Smart Playlists, you give iTunes certain parameters and it makes a playlist with songs that fit that criteria. (See Chapter 3.)

New Folder

The number of Playlists can get unwieldy and may require scrolling up and down to find your favorite mix. New Folder creates a nameable folder to help you organize your mix collections.

Add to Library ...

Use this option to find specific music files or folders
on your computer and tell iTunes to make them
accessible through your iTunes library—handy if
you imported music files before you started using
iTunes.

Close Window

Close the current window. Note this does not close
iTunes entirely.

Import ...

If you're using other digital music programs, use
this option to transfer music information to iTunes.

Export Song List (PC)

Use this option to have iTunes create a text listing of all the songs within the highlighted playlist.
Genre, time, and other details are included for each
song.

Export ... (Mac)

Use this option to have iTunes create a text listing of all the songs within the highlighted playlist.
Genre, time, and other details are included for each
song.

Export Library ...

Choose this to have iTunes create a complex website file called XML with information on every song in your library.

Back Up to Disc ...

Use this to back up your music collection with the touch of a button.

Get Info

Use this to get the details on the currently highlighted song, album, show, movie, or podcast. (See Chapter 3.)

My Rating

Adjust or create a 5-star rating for the highlighted song. (See Chapter 3.)

Edit Smart Playlist

With this, you can modify the criteria of the highlighted Smart Playlist. (See Chapter 3.)

Open in Windows Explorer (PC)

Windows Explorer is Microsoft's popular PC Internet browser. With this selection, iTunes opens the web browser.

Show in Finder (Mac)

Finder is a program that shows the location of all the files on your computer. This selection gives the location of the actual song file.

Show Current Song

If you're playing a song, use this to have iTunes open your computer directory and take you to where the song file is stored on your hard drive. This is useful when you're trying to locate the actual file.

Burn Playlist to Disc

If you have a blank, writable CD in your computer, select this to burn your current playlist to the disc. This is the same as the Burn Disc icon in the lower right corner. (See Chapter 3.)

Create an iMix ...

iTunes lets you post your favorite playlist of songs online through the iTunes Store so your friends and family can create the song list themselves. Use this option to create your own iMixes.

Sync iPod

This option moves your iTunes music to your iPhone/iPod. iTunes transfers as much music as your iPod can hold.

Transfer Purchases from iPod (Mac)

Use this option to move any music you've bought on the iTunes Store to the current computer.

Page Setup ...

Use this to set up the parameters for printing your iTunes screen.

Print ... (Mac)

Print the CD cover (displayed in the Album Artwork section in the lower-left corner) as well as track and artist information.

Exit (PC)

Shut down iTunes.

Edit

Here you can cut and paste songs, show duplicates, and set your iTunes preferences on PCs (see "Preferences ..." for Mac preferences). The first set of options are very similar to those in word processing programs like Microsoft Word.

Undo

Use this to undo the last action.

Cut

Cut the current selection.

Copy

Copy the current selection.

Paste

Place the previously cut or copied selection. This also pastes the previously cut or copied song.

Delete (Mac)

Delete the current selection.

Select All

Use this to highlight all the songs within the current list.

Select None

Use this to deselect any highlighted items.

Preferences (PC)

This brings up the powerful Preferences menu, where you can control your iPod options, CD burning techniques, and more.

Special Characters ... (Mac)

Prompts the Special Characters box, which allows you to add accents, umlauts, and other unusual items to your song or video names.

Controls

From this pull-down menu, you can control how your music plays. (See Chapter 3.)

Play

Use this to start the current selection. This is the same as the Play button located in the upper-left corner. (See Chapter 3.)

Next

Choose this to go to the next selection. It's the same as the Next button located in the upper-left corner. (See Chapter 3.)

Previous

This is a bit of a misnomer. Previous "rewinds" the current selection back to the beginning. Press it twice to go back to the previous selection. This is the same as the Previous button located in the upper-left corner. (See Chapter 3.)

Next Chapter

When listening to an audiobook, this moves you to the next chapter.

Previous Chapter

When listening to an audiobook, this moves you to the previous chapter.

Shuffle

This option randomizes the play order of your library or current playlist. It's the same as the Shuffle icon, the second icon in the lower-left corner. (See Chapter 3.)

Repeat Off

Choose this when you don't want iTunes to repeat the library or current playlist when it ends. It's the same as the Repeat icon, the third icon in the lower-left corner. (See Chapter 3.)

Repeat All

Check this when you want iTunes to start playing the library or current playlist again when you reach the end. It's the same as the Repeat icon, the third icon in the lower-left corner. (See Chapter 3.)

Repeat One

Check this to have iTunes repeat the current song until you say otherwise. It's the same as the Repeat icon, the third icon in the lower-left corner. (See Chapter 3.)

Volume Up

This increases the volume. It's the same as the volume bar in the upper-left corner. (See Chapter 3.)

Volume Down

This decreases the volume. It's the same as the volume bar in the upper-left corner. (See Chapter 3.)

Mute

Need to grab dinner? Answer the phone? Feed the cat? Use this to mute the music. (You'll still hear the usual sounds from your computer.)

Eject Disc

Use this to eject the current CD, DVD, iPod, or iPhone from your computer. It's the same as the Eject Disc icon in the lower-right corner. (See Chapter 3.)

View

Use this menu to control how your iTunes screen appears and at what size.

Show/Hide Browser

This option shows or removes the Browse area above your song list. It's the same as the Browse icon in the lower-right corner. (See Chapter 3.)

Show/Hide Artwork

This option shows or removes the album artwork from the lower-left corner of the screen. It does not delete the artwork. (See Chapter 3.)

Show/Hide MiniStore

Use this to show or hide the MiniStore at the bottom of the screen. (See Chapter 5.)

Show/Hide Equalizer (PC)

Use this to show or hide the music equalizer.

Turn On Visualizer (Mac)

The equivalent of a screen saver, the iTunes Visualizer creates a colorful light show on your screen, and the lights pulsate to the music. Press Escape to end the show.

Visualizer

Have another cool screensaver on your computer? Use this option to select another visualizer to display when you play music.

List View

The default setup, List View shows all your music, videos, audiobooks, and movies in alphabetical order. It is text only.

Album View

A spacious setup, Album View shows all your music, videos, audiobooks, and movies in alphabetical order along with the appropriate covers.

Cover Flow View

The most attractive setup, Cover Flow View magnifies the album covers, along with the song and album name, so you can flip through them like a jukebox. This setup is best for faster computers.

Half Size

This option makes the playing video appear in a small box in the middle of your screen surrounded by black. This mode is best for less-powerful computers because the video can slow down when it's bigger on the screen.

Actual Size

This option makes the playing video appear in a medium box in the middle of your screen surrounded by black. This mode is best for the average computer because the video can slow down when it's bigger on the screen.

Double Size

This option makes the playing video appear in a large box in the middle of your screen surrounded by a small black border. This mode is best for strong computers because the video can slow down on weaker computers.

Fit to Screen

This option fits the playing video into a very large box as wide as your screen surrounded by a small black border. This mode is best for strong computers because the video can slow down on weaker computers.

Full Screen

This option makes the playing video fill your entire screen. This mode is best for strong computers because the video can slow down on weaker computers.

Show Duplicates

This option shows any songs you have listed twice (or more) in your collection. You can delete the duplicates to create more space.

View Options ...

Use this to control how your music collection is listed.

Store

You can tailor your iTunes Store experience by modifying the options found here.

Search ...

Looking for a specific song, album, artist, movie, or TV show? Here's a quick way to find it.

Home

This takes you to the iTunes Store front page.

Previous Page

This moves you to the previous iTunes Store page (if available).

Next Page

This moves you to the next iTunes Store page (if available).

Check for Purchases ...

Even the most observant purchaser sometimes forget to actually download what he or she bought. Use this option to have iTunes check for any undownloaded purchases.

Authorize Computer ...

Use this to authorize the current computer to play music bought on your account.

Deauthorize Computer ...

Use this to stop the current computer from playing music bought on your account.

Sign Out

This option logs you out of your account to prevent any more purchases.

View My Account

Get the details of your account (if you're logged in) with this option.

Window (Mac)

For the Mac, Window controls the size of your iTunes player.

Minimize

Use this to minimize iTunes. It will still continue to play.

Zoom

This option toggles iTunes between full size and its small mode.

iTunes

Use this to make the iTunes player the dominant window always.

Equalizer

Use this to show or hide the music equalizer.

Bring All to Front

This option highlights and opens all the iTunes windows.

Party Shuffle

This opens Party Shuffle.

Advanced

Here you'll find a higher level of iTunes control, including radio programming, music authorizations, and conversion.

Switch to Mini Player (PC)

This option turns your iTunes display into a mini-player consisting of play controls and song information.

Open Stream ...

Here you can plug in the web addresses of particular radio programs you'd like to listen to.

Subscribe to Podcast ...

Find a particular podcast and subscribe to it with this option.

Convert Selection to AAC (PC)

AAC is a special music format only playable on Apple products like the iPhone and iPod. Use this option to have iTunes convert the selected file to ACC.

Convert Selection to MP3 (Mac)

MP3 is the most common music format. The iPhone and iPod use AAC, a special music format unique to Apple. This option converts the selected file into the more common format.

Convert ID3 Tags ...

Music information sometimes comes out wrong, particularly if it was imported to your computer by a program other than iTunes. This information is stored in something called an ID3 tag. Highlight the song, select this option, and choose Reverse Unicode. The problem will probably be fixed.

Consolidate Library ...

If your music is spread across different areas of your computer, use this option to have iTunes put your collection all in one particular file area.

Get Album Artwork

Use this to have iTunes find album artwork for your collection if it's available on the iTunes Store.

Get CD Track Names

Have iTunes find song data on a commercial CD using this option. You need to have an Internet connection for this to work. (See Chapter 3.)

Submit CD Track Names

Some commercial CDs are too obscure for iTunes to have the song information available. However, you may know the song title, composer, and such. This option allows you to type in the song information and give it to iTunes so other Apple users can benefit from your knowledge.

Join CD Tracks

Use this to remove the play gaps that usually occur between CD tracks. Highlight the two songs, and select this option.

Deauthorize Audible Account ... (Mac)

Apple has a special relationship with Audible.com, a distributor of audio books. There are a limited number of computers you can allow to play the books. This option allows you to deauthorize a computer so you may play it on other computers.

Help

Apple provides extensive help documents for iTunes. Some require an Internet connection.

iTunes Help

Here you can search a help database about iTunes and the iTunes Store. Type in a particular topic, such as Burning CDs, and it gives you all related tips and tricks. You can also print the information.

iTunes Service and Support

This takes you to the Apple Help website. It requires an Internet connection.

Keyboard Shortcuts

Check here for a listing of all the keyboard shortcuts you can use instead of pulling down menus or pressing icons.

iPod Help

Here you can search a help database about your iPod. Type in a particular topic, such as Transferring Music, and it gives you all related tips and tricks. You can also print the information.

iPod Service and Support

This takes you to the Apple Help website. It requires an Internet connection.

Apple TV Help

Here you can search a help database about your Apple TV. Type in a particular topic, such as Viewing Videos, and it gives you all related tips and tricks. You can also print the information.

iTunes Hot Tips (PC)

This option takes you to an Apple website with pointers on how to make the best of your iTunes setup. You need Internet access to view.

Shop for iTunes Products (PC)

This takes you to the iPod Accessories website, where you can purchase the latest iPod products. You need Internet access to enter. (See Chapter 8.)

Provide iTunes Feedback (PC)

Here you can communicate with Apple. Let them know about a CD you think should be available through the iTunes Store, about an error in the music listing, and other kinds of feedback. Apple does not reply in most cases, but the company says all feedback is noted. You need Internet access to talk with Apple.

Check for Updates (PC)

Use this to touch base with Apple to check for recent iTunes software updates. Apple usually sends you a notice when a new iTunes update is available, but this is convenient if you happen to miss the message. Internet access is required.

Run Diagnostics (PC)

Use this to have iTunes check the integrity of the current CD or DVD you have in your computer. It gives a short report after a brief spin of the CD drive.

About iTunes (PC)

Copyright information.

Index

T-U

Other "cool gadget" titles by Damon Brown